ESTIMATOR'S
EQUIPMENT INSTALLATION
MAN-HOUR MANUAL

T H I R D
E D I T I O N

Man-Hour Manuals and Other Books by John S. Page

Estimator's Electrical Man-Hour Manual/3rd Edition

Estimator's Equipment Installation
Man-Hour Manual/3rd Edition

Estimator's General Construction
Man-Hour Manual/2nd Edition

Estimator's Man-Hour Manual on Heating,
Air Conditioning, Ventilating, and Plumbing/2nd Edition

Estimator's Piping Man-Hour Manual
5th Edition

Conceptual Cost Estimating Manual

Cost Estimating Manual for
Pipelines and Marine Structures

John S. Page has wide experience in cost and labor estimating, having worked for some of the largest construction firms in the world. He has made and assembled numerous types of estimates including lump-sum, hard-priced, and scope, and has conducted many time and method studies in the field and in fabricating shops. Mr. Page has a B.S. in civil engineering from the University of Arkansas and received the Award of Merit from the American Association of Cost Engineers in recognition of outstanding service and cost engineering.

ESTIMATOR'S EQUIPMENT INSTALLATION MAN-HOUR MANUAL

T H I R D E D I T I O N

JOHN S. PAGE

G | P
P | ♥

Gulf Professional Publishing
An Imprint of Elsevier
Houston, Texas

To my two daughters
Patti and Terry

Estimator's Equipment Installation
Man-Hour Manual

Third Edition

Permissions may be sought directly from Elsevier's Science and Technology Rights Department in
Oxford, UK. Phone: (44) 1865 843830, Fax: (44) 1865 853333, e-mail: permissions@elsevier.co.uk.
You may also complete your request on-line via the Elsevier homepage: http://www.elsevier.com by
selecting "Customer Support" and then "Obtaining Permissions".

Gulf Professional Publishing
An Imprint of Elsevier
Book Division
P.O. Box 2608
Houston, Texas 77252-2608

Library of Congress Cataloging-in-Publication Data

Page, John S.
     Estimator's equipment installation man-hour manual / by John S.
Page. — 3rd ed.
          p.   cm.
     Includes bibliographical references and index.
     ISBN-13: 978-0-88415-287-3 ISBN-10: 0-88415-287-1 (alk. paper)
     1. Installation of industrial equipment—Estimates.   I. Title.
TS191.3.P33   1999
658.2'7—dc21   99-18582
                                                                        CIP

ISBN-13: 978-0-88415-287-3
ISBN-10: 0-88415-287-1

Transferred to Digital Printing 2009

Printed on acid-free paper (∞).

CONTENTS

Section 1 — EQUIPMENT

Section 2 — RELATED EQUIPMENT ITEMS

Section 3 — TECHNICAL INFORMATION

PREFACE

This third edition contains 26 new tables that cover a wide range of equipment installation activities. However, it is not the intent of this manual to produce anything new for the well-seasoned mechanical estimator whose ability, know-how and knowledge in this field are the products of years of schooling, actual competitive bidding, hard knocks, and time consuming analyzation of both good and bad estimates. Its main intention is to assist the partially experienced mechanical estimator by affording a basis for arriving at a reasonable dollar value for direct labor operations.

The many manhour tables that follow are the product of thousands of dollars spent for time studies and research analysis in this field. We believe that it will decrease the chance of error and allow the partially experienced estimator a greater advantage to more accurately determine the actual direct labor cost for the complete installation of mechanical equipment for a given industrial or chemical plant.

After careful analysis of these many studies we found that a productivity of 70% was a fair average for all crafts that might be involved in a normal mechanical contract. The direct labor manhours throughout this manual are based on this percentage.

You will find no cost as to material, equipment usage, warehousing and storage, fabricating shop set-up or overhead. If a material takeoff is available, this cost can be obtained at current prices from vendors who are to furnish the materials. Warehousing and storage, fabricating shop setup, equipment usage and overhead can readily be obtained by a good estimator who can visualize and consider the job schedule, size, and location. These are items which can and must be considered for the individual project.

Before an attempt is made to apply the following direct labor manhour tables we caution the estimator to be thoroughly familiar with the introduction on the following pages entitled, "Production and Composite Rate," which is the key to this type of estimating.

The Human Factor in Estimating

In this high-tech world of sophisticated software packages, including several for labor and cost estimating, you might wonder what a collection of man-hour tables offers that a computer program does not. The answer is the *human factor*. In preparing a complete estimate for a refinery, petrochemical, or other heavy industrial project one often confronts 12–18 major accounts, and each account has 5–100 or more sub-accounts, depending on the project and its engineering design. While it would seem that such numerous variables provide the perfect opportunity for computerized algorithmic solution, accurate, cost-effective, realistic estimating is still largely a function of human insight and expertise. Each project has unique aspects that still require the seasoned consideration of an experienced professional, such as general economy, projects supervision, labor relations, job conditions, construction equipment, and weather, to name a few.

Computers are wonderful tools. They can solve problems as no human can, but I do not believe construction estimating is their forté. I have reviewed several construction estimating software packages and have yet to find one that I would completely rely on. Construction estimating is an art, a science, and a craft, and I recommend that it be done by those who understand and appreciate all three of these facets. This manual is intended for those individuals.

John S. Page

INTRODUCTION
Production and Composite Rate

Herein lies a method for the application of the many manhour tables that follow.

Before one begins to think in terms of labor dollars for an estimate there are many things that must be considered. The most important of these is what we call productivity efficiency coupled with production elements. This is a must if the many manhour tables that follow are to be correctly applied and these items must be considered for each individual project.

After comparison of many projects, constructed under varied conditions, we have found that production elements can be grouped into six different classifications and that production percentages can be classified into five different categories.

The six different classifications of production elements are:

1. General Economy
2. Project Supervision
3. Labor Relations
4. Job Conditions
5. Equipment
6. Weather

The five ranges of productivity efficiency percentages are:

Type	Range
1. Very low	10-40%
2. Low	41-60%
3. Average	61-80%
4. Very Good	81-90%
5. Excellent	91-100%

Since there is such a wide range between the productivity percentages, let us attempt to evaluate each of the six elements, giving an example with each, and see just how a true productivity percentage can be obtained.

1. GENERAL ECONOMY: This is nothing more than the state of the nation or area in which your project is to be constructed. The things that should be reviewed and evaluated under this category are:
   a. Business trends and outlooks
   b. Construction volume
   c. Employment situation

Let us assume that after giving due consideration to these items you find them to be very good or excellent. This sounds good, but actually it means that your productivity range will be very low. This is due to the fact that with business being excellent the top supervision and craftsmen will be mostly employed and all that you will have to draw from will be inexperienced personnel. Because of this, in all probability, it will tend to create bad relationship between owner representatives, contract supervision, and the various craftsmen, thus making very unfavorable job conditions. On the other hand, after giving consideration to this element you may find the general economy to be of a fairly good average. Should this be the case, you should find that productivity efficiency tends to rise. This is due to the fact that under normal conditions there are enough good supervisors and craftsmen to go around, they are satisfied, thus creating good job conditions and understanding for all concerned. We have found, in the past, that general economy of the nation or area where your project is to be constructed, sets off a chain reaction to the other five elements. We, therefore, suggest that very careful consideration be given this item.

As an example, to show how a final productivity efficiency percentage can be arrived at, let us say that we are estimating a project in a given area and after careful consideration of this element, we find it to be of a high average. Since it is of a high average, but by no means excellent, we estimate our productivity percentage at 75%.

2. PROJECT SUPERVISION: What is the caliber of your supervision? Are they well-seasoned and experienced? What can you afford to pay them? What supply do you have to draw from? Things that should be looked at and evaluated under this element are:

        a. Experience
        b. Supply
        c. Pay

Like general economy this too must be carefully analyzed. If business is normal, in all probability, you will be able to obtain good supervision, but if business is excellent the chances are that you will have a poor lot to draw from. Should the contractor try to cut overhead by the use of cheap supervision he will usually wind up doing a very poor job. This usually results in a dissatisfied client, a loss of profit, and a loss of future work. This, like the attachment of the fee for a project, is a problem over which the estimator has no control. It must be left to management. All the estimator can do is to evaluate and estimate his project accordingly.

To follow through with our example, after careful analysis of the three items listed under this element, let us say that we have found our supervision will be normal for the project involved and we arrive at an estimated productivity rate of 70%.

3. LABOR CONDITIONS: Does your organization possess a good labor relations man? Are there experienced first class satisfied craftsmen in the area where your project is to be located? Like project supervision, things that should be analyzed under this element are:

> a. Experience
> b. Supply
> c. Pay

A check in the general area where your project is to be located should be made to determine if the proper experienced craftsmen are available locally, or will you have to rely on travelers to fill your needs. Can and will your organization pay the prevailing wage rates?

For our example, let us say that for our project we have found our labor relations to be fair but feel that they could be a little better and that we will have to rely partially on travelers. Since this is the case, we arrive at an efficiency rating of 65% for this element.

4. JOB CONDITIONS: What is the scope of your project and just what work is involved in the job? Will the schedule be tight and hard to meet, or will you have ample time to complete the project? What kind of shape or condition is the site in? Is it low and mucky and hard to drain, or is it high and dry and easy to drain? Will you be working around a plant already in production? Will there be tie-ins, making it necessary to shut down various systems of the plant? What will be the relationship between production personnel and construction personnel? Will most of your operations be manual, or mechanized? What kind of material procurement will you have? There are many items that could be considered here, dependent on the project; however, we feel that the most important items that should be analyzed under this element are as follows:

> a. Scope of work
> b. Site conditions
> c. Material procurement
> d. Manual and mechanized operations

By a site visitation and discussion with owner representatives, coupled with careful study and analysis of the plans and specifications, you should be able to correctly estimate a productivity percentage for this item.

For our example, let us say that the project we are estimating is a completely new plant and that we have ample time to complete the project but the site location is low and muddy. Therefore, after evaluation we estimate a productivity rating of only 60%.

5. EQUIPMENT: Do you have ample equipment to complete your project? Just what kind of shape is it in and will you have good maintenance and repair help? The main items to study under this element are:

    a. Usability
    b. Condition
    c. Maintenance and repair

This should be the simplest of all elements to analyze. Every estimator should know what type and kind of equipment his company has, as well as what kind of mechanical shape it is in. If equipment is to be obtained on a rental basis then the estimator should know the agency he intends to use as to whether they will furnish good equipment and good maintenance.

Let us assume for our example, that our company equipment is in very good shape, that we have an ample supply to draw from and that we have average mechanics. Since this is the case we estimate a productivity percentage of 70%.

6. WEATHER: Check the past weather conditions for the area in which your project is to be located. During the months that your company will be constructing, what are the weather predictions based on these past reports? Will there be much rain or snow? Will it be hot and mucky or cold and damp? The main items to check and analyze here are as follows:

    a. Past weather reports
    b. Rain or snow
    c. Hot or cold

This is one of the worst of all elements to be considered. At best all you have is a guess. However, by giving due consideration to the items as outlined under this element, your guess will at least be based on past occurrences.

For our example, let us assume that the weather is about half good and half bad during the period that our project is to be constructed. We must then assume a productivity range of 50% for this element.

We have now considered and analyzed all six elements and in the examples for each individual element have arrived at a productivity efficiency percentage. Let us now group these percentages together and arrive at a total percentage:

Item	Productivity Percentage
1. General Economy	75
2. Production Supervision	70

3. Labor Relations	65
4. Job Conditions	60
5. Equipment	70
6. Weather	50
Total	390%

Since there are six elements involved, we must now divide the total percentage by the number of elements to arrive at an average percentage of productivity.

$$390\% \div 6 = 65\% \text{ average productivity efficiency}$$

At this point we caution the estimator. This example has been included as a guide to show one method that may be used to arrive at a productivity percentage. The preceding elements can and must be considered for each individual project. By so doing, coupled with the proper manhour tables that follow, a good labor value estimate can be properly executed for any place in the world, regardless of its geographical location and whether it be today or 20 years from now.

Next we must consider the composite rate. In order to correctly arrive at a total direct labor cost, using the manhours as appear in the following tables, this must be done.

Most organizations consider field personnel with a rating of superintendent or greater as a part of job overhead, and that of general foreman or lower as direct job labor cost. The direct manhours as appear on the following pages have been determined on this basis. Therefore, a composite rate should be used when converting the manhours to direct labor dollars.

Again the estimator must consider labor conditions in the area where the project is to be located. He must ask himself how many people will he be allowed to use in a crew, can he use crews with mixed crafts, and how many crews of the various crafts will be need.

In the following example that may be used to obtain a composite rate, we assume that a certain project has a certain piece of equipment to be installed and that a mixed crew consisting of the following crafts will be needed:

Assumed Crew — Productive Hours:

Pipefitters	4 each for 8 hours each =	32 hours
Millwrights	2 each for 4 hours each =	8 hours
Electricians	2 each for 2 hours each =	4 hours
Truck Driver	1 each for 1 hours each =	1 hours
Total		45 hours

Assumed rate of craft in the given area:

Craft	Foreman	Journeyman
Pipefitters	$23.50	$23.00
Millwrights	$21.00	$20.50
Electricians	$21.50	$21.00
Truck Drivers	—	$13.00

Assume that foreman are dead weight for the hypothetical case since they will not be working with their tools. Because of their supervisory capacity their time must be considered and charged to the crew.

Assumed crew for composite rate:
Foreman:

Pipefitter	4 hours	@ $23.50 = $94.00
Millwright	2 hours	@ $21.00 = $42.00
Electrician	1 hour	@ $21.50 = $21.50

Journeyman:

Pipefitter	32 hours	@ $23.00 =	$736.00
Millwright	8 hours	@ $20.50 =	$164.00
Electrician	4 hours	@ $21.00 =	$ 84.00
Truck Driver	1 hour	@ $13.00 =	$ 13.00
			$1,154.50

$1,154.50 ÷ 45 Productive hours = $25.65 Composite rate for 100% time.

It is well to note, at this time, that as was stated in the preface to this manual the manhours are based on an average productivity of 70% for all crafts involved. Therefore, the composite rate of $25.65 as figured, becomes equal to 70% productivity.

Let us now assume that we have evaluated a certain project to be bid and find it to be of a low average with an overall productivity rating of only 65%. This means a loss of 5% of time paid for manhour. Therefore, the composite rate should have an adjustment of 5% as follows:

$$\$25.65 \times 105\% = \$26.93 \text{ (Composite rate for 65\% productivity)}$$

Simply by multiplying the number of manhours estimated for a given block or item of work by the arrived at composite rate, a total estimated direct labor cost, in dollar value can be easily and accurately obtained.

It is our express desire and sincere hope that the foregoing will enable the ordinary mechanical estimator to turn out a better labor estimate and assist in the elimination of much guesswork.

Section 1

EQUIPMENT

It is the intent of this section to cover as nearly as possible all required operations for the installation of individual pieces of process and other equipment as may be required for a process or industrial plant.

The manhours listed are for direct labor only and have no bearing on equipment or material costs, construction equipment rental, small tools, or overhead. These are items that must be given consideration for the individual project if a complete equipment installation estimate is to be obtained.

All labor for unloading from railroad cars or trucks, storing in storage yard or warehouse, hauling to erection site and installing have been given due consideration in the manhours listed.

Before attempting to apply the manhours in this or any section, we caution the estimator to be thoroughly familiar with the introduction to the manual.

PACKAGED STEAM BOILERS

Saturated Steam — 100 to 235 PSIG

MANHOURS REQUIRED EACH

BOILER				FORCED DRAFT FAN	
Capacity Pounds Per Hour	Boiler Weight in Pounds	Firing & Control Equipment Weight in Pounds	Manhours	Fan Horsepower	Manhours
30,000	51,000	6,700	170.0	20	50.0
35,000	54,000	7,000	180.0	20	50.0
40,000	58,000	7,200	185.0	25	62.5
45,000	61,000	7,400	190.0	25	62.5
50,000	64,000	7,400	196.0	30	75.0
60,000	71,000	7,800	201.0	40	100.0
75,000	89,000	10,800	204.0	60	120.0
80,000	83,000	12,200	220.0	60	120.0
90,000	91,000	12,500	264.0	100	150.0
100,000	98,000	12,800	286.0	100	150.0
110,000	88,000	13,100	260.0	125	187.5
115,000	95,000	15,000	312.0	125	187.5
125,000	97,000	16,300	338.0	150	225.0
135,000	100,000	21,300	364.0	150	225.0
150,000	103,000	24,600	380.0	200	300.0

Manhours include unloading, handling, job hauling up to 2000 feet, rigging, picking, setting, aligning, and checking out of item as outlined.

Manhours exclude installation of piping and electrical items and hook-up.

Boiler manhours are for installation of complete boiler units including firing and control equipment.

Forced draft fan manhours are for complete installation and hook-up of fans.

WATER HEATING BOILERS—
PACKAGE TYPE ELECTRIC HYDRONIC BOILERS

MANHOURS REQUIRED EACH

Output BTU/Hour	Installation Manhours
34,000	4.0
51,000	5.5
68,000	7.0
81,000	8.0

CAST IRON GAS-FIRED BOILERS

MANHOURS REQUIRED EACH

Net IBR Rating MBH	Installation Manhours
46.1	13.2
68.7	17.6
104.3	22.0
156.5	27.4
208.7	36.5
260.9	41.0
313.0	45.6
365.8	47.9
417.4	52.8
469.6	57.6
556.5	70.2
695.7	75.4
834.8	93.0
973.9	96.0

Manhours include unloading, handling, job hauling up to 2000 feet, rigging, picking, setting, aligning and checking out of boiler as outlined.

Manhours exclude installation of gas, steam, water or other piping or installation of motor, starter and power wiring.

Electric hydronic boilers are complete packages including automatic temperature controls and are pre-wired for connection to 240 volt power source.

Cast iron gas fired boilers include automatic gas valve, fail safe type pilot, manual shut-off, drain cock, low water cut-off, safety valve, gauge glass, steam pressure gauge and draft diverter.

STEEL BOILER STACKS

MANHOURS PER LINEAR FOOT

Stack Diameter Inches	Plate Thickness Inches	Weight Per Foot Pounds	Manhours Per Linear Foot
24	1/4	65	0.46
30	1/4	81	0.57
36	1/4	96	0.68
42	1/4	112	0.78
42	5/16	140	0.98
48	1/4	128	0.90
48	5/16	160	1.12
54	1/4	144	1.00
54	5/16	180	1.26
54	3/8	216	1.51
60	1/4	160	1.12
60	5/16	200	1.40
60	3/8	240	1.68

Manhours include unloading, handling, job hauling up to 2000 feet, rigging, picking, setting, aligning, bolting and guying stack up to 50 feet long in position.

Manhours exclude make-up of joints or field add-ons.

FIRED HEATERS

MANHOURS EACH

Approximate Height	Approximate Weight Pounds	Manhours
5'0"	1,100	55.0
5'6"	1,250	62.5
8'0"	1,250	62.5
9'0"	1,500	75.0
10'0"	1,950	97.5
10'6"	3,350	134.0
11'0"	2,000	100.0
12'0"	2,600	104.0
13'0"	5,950	150.0
14'0"	8,850	177.0
15'6"	5,500	140.0
18'6"	7,150	165.0
21'6"	8,800	177.0
28'6"	16,000	240.0
37'6"	29,650	300.0

Manhours include unloading, handling, job hauling up to 2000 feet, rigging, picking, setting, aligning, and checking out of heater as outlined.

Manhours exclude installation of piping, electrical circuit and connections.

Manhours are for installation of shop assembled heater, centrifugal air blower, and control panel.

Heaters are of the gas, oil, or dual-fired type.

DEAERATING HEATERS

MANHOURS REQUIRED EACH

Size Diameter x Length	Approximate Weight Pounds	Outlet Capacity Pounds Per Hour	Manhours
4'0" x 3'0"	930	2,800	30.0
4'0" x 3'0"	950	4,000	30.0
4'0" x 3'0"	970	6,000	30.0
4'0" x 3'6"	1,050	8,500	30.0
4'0" x 4'0"	1,100	10,000	30.0
4'0" x 5'0"	1,200	13,000	30.0
4'0" x 7'0"	1,500	16,000	30.0
5'0" x 4'6"	2,000	20,000	30.0
5'0" x 7'6"	2,500	30,000	34.0
5'0" x 10'0"	3,000	40,000	34.0
5'0" x 13'0"	3,500	50,000	38.0
6'0" x 7'6"	4,500	60,000	40.0
6'0" x 8'9"	5,000	70,000	60.0
6'0" x 10'0"	5,500	80,000	60.0
6'0" x 11'6"	6,000	90,000	60.0
6'0" x 13'0"	6,500	100,000	64.0
7'0" x 11'6"	7,500	140,000	70.0
7'0" x 15'0"	8,900	180,000	70.0
8'0" x 15'0"	11,000	240,000	90.0
8'0" x 19'0"	13,000	300,000	90.0
8'0" x 23'0"	14,500	360,000	90.0

Manhours include unloading, handling, job hauling up to 2000 feet, rigging, picking, setting, aligning, and checking out of heater as outlined.

Manhours exclude installation of piping and connections.

Manhours are for installation of two-stage, spray-type heater with internal vent condenser.

TANK HEATING COILS

MANHOURS EACH

Heating Surface Square Feet	Manhours Each
250	29
265	32
498	38

FIRED IMMERSION HEATERS

MANHOURS EACH

Tube Size Inches	Capacity BTU/Hour	Heating Surface Square Feet	Manhours Each
8	750,000	344	58
10	1,250,000	576	62
12	1,800,000	860	67

Manhours include unloading, handling, job hauling up to 2000 feet, installing, and inspecting.

Tank heating coils manhours are based on all carbon steel construction.

Fired immersion heater manhours are based on installing factory skid mounted units.

All manhours exclude installation of piping, electrical circuits, and connections.

CLASSIFICATION EQUIPMENT

MANHOURS REQUIRED EACH

Item Description	Approximate Weight Pounds	Manhours Required
Cone Type Pelletizers		
Cone Size — 4'0"	2,100	24
Cone Size — 8'0"	15,000	36
Cone Size — 12'0"	25,500	48
Cone Size — 16'0"	33,000	60
Pressure Sifter		
48" Diameter — 3-Screen	1,300	16
Rotary Screens		
Frame 15" x 28" — Cloth Area 2 to 6 SF.	400	8
Frame 30" x 30" — Cloth Area 4 to 16 SF.	450	8
Frame 30" x 45" — Cloth Area 8 to 16 SF.	500	10
Frame 40" x 40" — Cloth Area 8 to 48 SF.	1300/1900	24
Frame 40" x 60" — Cloth Area 14 to 70 SF.	1,600/2,050	24
Frame 40" x 80" — Cloth Area 19 to 76 SF.	1,700/2,050	24
Turbo-Screen Classifiers		
4'2" High x 3'4" Wide x 3'2" Deep	575	10
6'4" High x 3'8" Wide x 3'9" Deep	1,300	20

Manhours include unloading, handling, job hauling up to 2000 feet, rigging, picking, setting, aligning, and checking out of complete item as outlined above.

Manhours exclude installation of electrical circuits and connections.

The number of sieves installed in the above rotary screens will determine the actual screen area capacity.

Turbo-screen classifiers, are complete units including cyclones and screens.

COMPRESSORS—RECIPROCATING
Gas Engine Driven

MANHOURS EACH

Compressor Horsepower	Approximate		Total Installation Manhours
	Total Weight Pounds	Heaviest Piece Pounds	
330	31,700	30,000	400.0
440	40,000	37,400	500.0
550	45,300	43,000	550.0
660	53,800	51,500	630.0
680	43,800	38,500	630.0
1,000	64,000	61,000	750.0
1,080	74,000	66,000	810.0
1,300	84,000	76,000	975.0
1,500	87,000	83,000	1,100.0
1,620	106,000	98,000	1,150.0
1,730	119,000	111,000	1,175.0
2,000	128,000	120,000	1,225.0
2,160	137,000	129,000	1,300.0
2,600	140,000	134,000	1,560.0
3,000	211,000	162,000	1,650.0
3,750	256,000	196,000	1,990.0
4,000	302,000	233,000	2,050.0
4,500	340,000	262,000	2,250.0
6,000	403,000	156,000	2,400.0

Manhours include unloading, handling, job hauling up to 2000 feet, uncrating of packages, rigging, picking, setting, aligning, and checking out of all packages or pieces.

Manhours do not include installation of starting air compressor, incoming or outgoing piping, or their connections.

COMPRESSORS—RECIPROCATING
Electric Motor Driven

MANHOURS EACH

Compressor Horsepower	Number of Cylinders	Approximate Weight		Manhours
		Total Less Motor Pounds	Heaviest Piece Pounds	
400	2	23,200	8,900	550.0
800	4	42,000	15,150	610.0
1,500	4	58,700	23,600	750.0
2,500	4	103,800	29,890	1,150.0
4,000	4	124,500	42,800	1,225.0
6,000	6	176,700	56,800	1,600.0
8,000	8	234,500	75,700	1,800.0
10,000	6	349,100	101,300	2,250.0
12,000	8	455,100	127,700	2,400.0

Manhours include unloading, handling, job hauling up to 2000 feet, uncrating of packages, rigging, picking, setting, aligning, and checking out of all packages or pieces.

Manhours do not include installation of incoming or outgoing piping, electrical circuits, or their connections.

AIR COMPRESSORS—CENTRIFUGAL
Packaged Units

MANHOURS EACH

Nominal Capacity CFM	Driver Horsepower	Approximate Weight Pounds	Manhours
1,250	300	10,600	48.0
1,500	350	10,850	48.0
2,100	450	12,000	60.0
2,500	600	12,300	60.0
3,000	700	23,500	96.0
3,500	800	24,000	96.0
4,000	900	25,000	108.0
5,000	1,000	25,400	108.0
6,000	1,250	47,000	144.0
7,000	1,500	48,000	144.0
8,000	1,750	48,800	144.0
11,000	2,500	72,000	192.0
15,000	3,500	90,000	240.0

Manhours include unloading, handling, job hauling up to 2000 feet, rigging, picking, setting, aligning and checking out of factory preassembled compressor units.

Manhours exclude installation of incoming or outgoing pipe, electrical circuits, and their connections.

AIR COMPRESSORS

Industrial Service
Air-Cooled, Two-Stage—60 To 250 PSI

MANHOURS EACH

Motor Horsepower	Approximate Weight Pounds	MANHOURS
25	2,050	50
40	2,400	80
50	2,825	100
75	3,900	112
100	4,425	125
125	4,625	125

Automotive Service
Air-Cooled—150 and 200 PSI
MANHOURS EACH

Motor Horsepower	Approximate Weight Pounds	MANHOURS	
		Single Stage	Two Stage
1/2	200	12	—
3/4	225	14	—
1	240	14	—
1-1/2	450	—	20
2	450	—	20
3	500	—	22
5	665	—	24
7-1/2	910	—	30
10	1,080	—	40
15	1,390	—	48

Manhours include unloading, handling, job hauling up to 2000 feet, rigging, picking, setting, aligning, and checking out of factory preassembled compressors as outlined.

Manhours exclude installation of incoming or outgoing piping, electrical circuits, and their connections.

PACKAGED AIR COMPRESSOR UNITS

Single-Stage—Crosshead Type

90 to 125 psig

MANHOURS EACH

Motor Horsepower	Approximate Weight Pounds	Manhours Each
Oil Lubricated		
20	1,950	58
25	2,400	61
30	2,800	64
40	3,800	67
40	3,800	67
50	3,900	70
60	5,100	86
60	7,000	91
75	7,000	91
100	8,100	95
100	7,500	93
125	8,600	96
Non-lubricated		
20	2,000	58
25	2,475	61
30	3,000	64
40	4,000	70
40	4,000	70
50	4,100	70
60	5,400	86
60	7,000	91
75	7,000	91
100	8,100	95
100	7,500	93
125	8,600	96

Manhours include unloading, handling, job hauling up to 2000 feet, rigging, picking, setting, aligning, and checking factory preassembled compressors as outlined.

Manhours exclude installation of incoming or outgoing piping, electric circuits and their connections.

TWO-STAGE AIR COMPRESSORS

MANHOURS EACH

Motor Horsepower	Approximate Weight Pounds	Manhours Each
150 to 300 psig		
30	2,700	67
50	4,200	71
100	7,900	105
150	10,900	115
300 to 500 psig		
30	2,700	67
40	4,200	71
100	7,850	105
150	10,900	115

Note: Compressors are water-cooled, factory-assembled units.

Manhours include unloading, handling, job hauling up to 2000 feet, rigging, picking, setting, aligning, and checking preassembled compressors as outlined.

Manhours exclude installation of incoming or outgoing piping, electrical circuits, and their connections.

AIR POWER COMPRESSORS

MANHOURS EACH

Motor	Manhours Each			
Horsepower	1	2	3	4
150	132	132	139	139
200	139	132	139	139
250	144	144	151	151
300	151	151	159	159
350	156	—	163	163
400	164	—	172	—
500	168	—	176	—
600	192	—	202	—

1. Lubricated type, two-stage, 80 to 150 psig.
2. Lubricated type, two-stage, 175 to 200 psig.
3. Lubricated type, rated for sea level to 3300 feet altitude, 125 psig maximum discharge pressure, synchronous motor driven.
4. Lubricated type, rated for sea level to 3300 feet altitude, 125 psig maximum discharge pressure, induction motor driven.

Manhours include unloading, handling, job hauling up to 2000 feet, rigging, picking, setting, aligning, and checking factory preassembled compressor as outlined.

Manhours exclude installation of incoming or outgoing piping, electrical circuits, and their connections.

INTEGRAL GAS ENGINE COMPRESSOR

MANHOURS EACH

Rated Brake Horsepower	Approximate Weight Pounds	Manhours Each
1,100	135,000	1,350
1,650	165,000	1,650
1,080	85,000	850
1,620	115,000	1,150
1,440	145,000	1,450
2,160	170,000	1,700
1,300	115,000	1,150
1,730	150,000	1,500
2,600	185,000	1,850
3,000	255,000	2,550
3,750	315,000	3,150
4,500	380,000	3,800
6,000	495,000	4,950

Manhours include unloading, handling, job hauling up to 2000 feet, rigging, picking, setting, aligning, and checking preassembled compressor and gas engine driver as outlined.

Manhours exclude installation of incoming or outgoing piping and their connections.

AIR DRYERS—REFRIGERATED TYPE

MANHOURS EACH

Capacity SCFM @ 100 PSIG	Approximate Weight Pounds	Size Length x Width x Height — Inches	Refrigeration Compressor Horsepower	MANHOURS
5	73	22 x 15 x 12	1/6	12.0
10	83	22 x 15 x 16	1/5	12.0
25	150	22 x 15 x 31	1/4	16.0
35	155	22 x 15 x 31	1/3	16.0
50	275	28 x 20 x 38	1/2	18.0
75	300	28 x 20 x 38	3/4	18.0
100	440	28 x 26 x 41	1	20.0
125	450	28 x 26 x 41	1	20.0
250	1,085	38 x 36 x 70	2	24.0
375	1,235	38 x 36 x 70	3	24.0
575	1,720	47 x 38 x 70	4	30.0
700	2,100	47 x 38 x 70	5	30.0
1,000	3,250	68 x 47 x 83	6	36.0
1,200	4,100	68 x 47 x 83	8	48.0
1,700	5,400	66 x 58 x 83	10	56.0

Manhours include unloading, handling, job hauling up to 2000 feet, rigging, picking, setting, aligning, and checking out of factory preassembled units as listed.

Manhours exclude installation of incoming or outgoing piping, electrical circuits, and their connections.

These dryers are of the refrigerated compressed air type used for removing moisture by cooling the air to a pressure dewpoint of 35°F, thereby removing water and lube oil which is discharged.

AIR DRYERS—CHILLER TYPE

MANHOURS EACH

Capacity SCFM @ 100 PSIG	Approximate Weight Pounds	Size Length x Width x Height — Inches	Refrigeration Compressor Horsepower	MANHOURS
2,500	7,000	142 x 68 x 96	10	56.0
3,750	8,500	162 x 68 x 96	15	56.0
5,000	9,000	177 x 68 x 96	20	56.0
6,250	11,500	177 x 68 x 96	25	68.0
7,500	13,500	188 x 84 x 105	30	68.0
10,000	15,750	200 x 84 x 105	40	72.0
12,500	19,000	200 x 84 x 105	50	81.0
15,000	20,500	200 x 84 x 105	60	90.0
18,750	23,500	246 x 90 x 109	75	94.0
25,000	27,750	264 x 90 x 109	100	100.0

Manhours include unloading, handling, job hauling up to to 2000 feet, rigging, picking, setting, aligning, and checking out of factory preassembled skid-mounted units as listed.

Manhours exclude installation of incoming and outgoing piping, electrical circuits, and their connections.

These chiller type air dryers dry compressed air by refrigeration, cooling the air to a low temperature which condenses the water vapor in the air.

AIR DRYERS—CHILLER TYPE
For Oil Free Service

MANHOURS EACH

Capacity SCFM @ 100 PSIG	Approximate Weight Pounds	Size Length x Width x Height – Inches	MANHOURS
2,980	6,000	139 x 70 x 81	56.0
4,655	7,500	141 x 70 x 81	56.0
6,515	8,000	141 x 70 x 82	56.0
7,448	10,500	152 x 77 x 89	64.0
10,240	13,500	181 x 77 x 93	68.0
13,965	14,700	183 x 79 x 96	70.0
16,760	18,000	195 x 89 x 100	76.0
20,480	19,500	196 x 95 x 104	81.0
26,070	22,000	206 x 107 x 112	90.0
26,790	26,000	206 x 111 x 115	96.0

Manhours include unloading, handling, job hauling up to 2000 feet, rigging, picking, setting, aligning, and checking out of factory preassembled skid-mounted units as listed.

Manhours exclude installation of incoming and outgoing piping, electrical circuits, and their connections.

These chiller type air dryers dry compressed air by refrigeration, cooling the air to a low temperature which condenses the water vapor in the air.

AIR DRYERS—DESICCANT TYPE

MANHOURS EACH

Size Inches			Weight Pounds	Connection Size	Manhours Each
Height	Width	Depth			
Wall-Mounted—2 to 50 SCFM					
25	20	7	40	¼"	10
25	20	7	40	¼"	10
25	20	7	50	¼"	12
35	25	10	140	½"	14
55	30	20	300	¾"	19
Floor-Mounted—90 to 1000 SCFM					
75	27	21	400	1"	29
89	27	21	440	1"	30
67	29	21	470	1"	32
77	29	21	520	1"	33
93	33	21	650	1½"	38
81	34	21	700	1½"	40
93	34	21	900	1½"	42
90	38	21	1,100	1½"	44
112	40	26	1,300	2"—150# flg.	48
94	45	28	1,500	2"—150# flg.	50
104	45	28	1,700	2"—150# flg.	53
98	45	28	1,900	2"—150# flg.	56
105	50	29	2,000	3"—150# flg.	58
113	50	29	3,000	3"—150# flg.	60
Floor-Mounted—1600 to 6,400 SCFM					
131	77	51	4,000	3"—150# flg.	67
134	89	52	5,500	4"—150# flg.	77
142	103	53	8,500	4"—150# flg.	96
151	124	61	12,000	6"—150# flg.	115
154	139	73	15,000	6"—150# flg.	121

Manhours include unloading, handling, job hauling up to 2000 feet, rigging, picking, setting, aligning, checking total dryer including checking calibration of preinstalled instruments.

Manhours exclude installation of main piping lines and electrical circuits.

AIR DRYER FILTERS AND AUTOMATIC DRAIN VALVES

MANHOURS EACH

Connection Size & Type	Maximum Flow SCFM	Cartridge Filters	Automatic Drain Valve	Manhours Each
¼" scwd.	20	1	1	4.2
½" scwd.	35	1	1	4.2
1" scwd.	60	1	1	4.2
1" scwd.	120	2	1	4.2
1" scwd.	180	3	1	4.2
1½" scwd.	350	1	1	4.8
2" scwd.	700	1	1	4.8
3" - 150# rf. flg.	1,050	3	1	7.2
3" - 150# rf. flg.	1,400	4	1	7.2
4" - 150# rf. flg.	2,100	3	1	9.6
4" - 150# rf. flg.	2,800	4	1	9.6
6" - 150# rf. flg.	4,200	6	1	12.0
6" - 150# rf. flg.	5,600	8	1	12.0

Manhours include unloading, handling, job hauling up to 2000 feet, rigging, picking, setting, aligning, and checking preassembled dryer filter units, and installation of automatic drain valves.

Manhours are for installation of either prefilter or afterfilter type cartridges as may be required.

Manhours exclude installation of incoming or outgoing piping and electrical circuits or their connections.

GAS PULSATION DAMPERS

MANHOURS EACH

Damper Size Diam. × Length	Manhours Each
6″ × 30″	24
8″ × 40″	29
10″ × 50″	31
12″ × 60″	33
14″ × 70″	34
16″ × 80″	36
18″ × 90″	38
20″ × 110″	40
24″ × 120″	42

Manhours include unloading, handling, job hauling up to 2000 feet, rigging, picking, setting, aligning and checking dampers as outlined.

Manhours exclude installation of piping and piping connections.

CONVEYORS—OPEN BELT

INSTALLATION HOURS

Length Linear Feet	18-inch Belt Width	24-inch Belt Width	MANHOURS 30-inch Belt Width	36-inch Belt Width	42-inch Belt Width
10	38.4	41.6	44.8	51.2	55.3
20	59.2	64.8	75.2	84.8	93.3
30	78.4	86.4	99.2	112.0	125.4
40	96.0	108.8	126.4	139.2	158.7
50	110.4	126.4	144.0	161.6	184.2
60	126.4	144.0	168.0	192.0	218.9
70	140.8	160.0	192.0	216.0	246.2
80	155.2	176.0	216.0	240.0	273.6
90	176.0	192.0	240.0	264.0	303.6
100	192.0	208.0	256.0	288.0	331.2
200	288.0	336.0	416.0	448.0	515.2
300	384.0	448.0	544.0	592.0	668.9
400	448.0	528.0	648.0	736.0	824.3
500	544.0	624.0	768.0	864.0	967.7
600	608.0	720.0	864.0	976.0	1,093.0
700	688.0	784.0	968.0	1,104.0	1,214.0
800	752.0	880.0	1,088	1,232.0	1,355.0
900	816.0	944.0	1,144.0	1,312.0	1,443.0
1,000	864.0	1,008.0	1,264.0	1,424.0	1,566.0

Manhours include unloading, job hauling up to 2000 feet, rigging, picking, setting, aligning, and checking out of conveyor and all components.

Manhours exclude installation of walkways, covers or electrical circuits, and connections.

Installation of "A" frames, truss complete with idlers, conveyor frame, head pulley and drive, fixed or screw take-up tail pulley, snub and bent pulleys, horizontal gravity take-up, vertical gravity take-up holdback, belting, and belt splicing are included in the manhours.

CONVEYORS—BELT ENCLOSED WITH WALKWAY

INSTALLATION MANHOURS

Length Linear Feet	MANHOURS				
	18-inch Belt Width	24-inch Belt Width	30-inch Belt Width	36-inch Belt Width	42-inch Belt Width
10	76.8	88.0	92.8	100.8	108.9
20	124.8	140.8	176.8	184.0	282.4
30	160.0	192.0	208.0	256.0	286.7
40	208.0	264.0	272.0	288.0	328.3
50	240.0	288.0	304.0	352.0	401.3
60	264.0	304.0	352.0	416.0	474.3
70	288.0	352.0	400.0	564.0	643.0
80	328.0	400.0	432.0	528.0	601.9
90	352.0	424.0	480.0	600.0	690.0
100	384.0	464.0	512.0	624.0	717.6
200	640.0	768.0	880.0	1,056.0	1,193.2
300	848.0	992.0	1,200.0	1,456.0	1,630.0
400	1,088.0	1,248.0	1,672.0	1,760.0	1,971.0
500	1,248.0	1,456.0	1,728.0	2,080.0	2,330.0
600	1,424.0	1,680.0	2,080.0	2,560.0	2,816.0
700	1,600.0	1,920.0	2,400.0	2,880.0	3,168.0
800	2,000.0	2,440.0	2,560.0	3,040.0	3,313.0
900	2,080.0	2,560.0	2,800.0	3,520.0	3,802.0
1,000	2,240.0	2,720.0	3,040.0	3,680.0	3,974.0

Manhours include unloading, job hauling up to 2000 feet, rigging, picking, setting, aligning, and checking out of conveyor and all components.

Manhours exclude installation of electrical circuits and connections.

Installation of "A" frames, truss complete with idlers, conveyor frame, head pulley and drive, fixed or screw take-up tail pulley, snub and bent pulleys, horizontal gravity take-up, vertical gravity take-up, holdback, belting, belt splicing, and steel framed walkway with angle hand rail, knee rail, toe plate and wooden walk, and metal belt enclosure are included in the manhours.

CONVEYORS–STEEL SCREW

INSTALLATION MANHOURS

Length Linear Feet	MANHOURS			
	6-inch Diameter	12-inch Diameter	16-inch Diameter	20-inch Diameter
5	28.8	40.0	62.4	88.0
6	32.0	43.2	67.2	97.6
7	35.2	46.4	73.6	105.6
8	38.4	49.6	80.0	113.6
9	41.6	54.4	86.2	121.6
10	43.2	56.0	92.8	128.0
20	59.2	83.2	136.0	192.0
30	72.0	105.6	192.0	256.0
40	83.2	126.4	224.0	288.0
50	94.4	142.4	240.0	320.0
60	102.4	158.4	256.0	352.0
70	112.0	192.0	272.0	376.0
80	120.0	216.0	296.0	416.0
90	128.0	240.0	312.0	448.0
100	134.4	272.0	336.0	464.0

Manhours include unloading, job hauling up to 2000 feet, rigging, picking, setting, aligning, and checking out of conveyor and all components.

Manhours exclude installation of electrical circuits and connections.

CONVEYORS–STAINLESS STEEL SCREW

INSTALLATION MANHOURS

Length Feet	MANHOURS			
	6-inch Diameter	12-inch Diameter	16-inch Diameter	20-inch Diameter
5	118.0	150.0	320.0	440.0
6	130.0	166.0	340.0	480.0
7	142.0	182.0	360.0	520.0
8	154.0	198.0	380.0	560.0
9	162.0	220.0	390.0	580.0
10	176.0	220.0	420.0	600.0
20	280.0	340.0	600.0	660.0
30	340.0	420.0	760.0	820.0
40	380.0	500.0	840.0	1000.0
50	440.0	560.0	980.0	1160.0
60	480.0	620.0	1060.0	1300.0
70	520.0	700.0	1140.0	1400.0
80	570.0	750.0	1240.0	1520.0
90	580.0	780.0	1320.0	1620.0
100	620.0	820.0	1380.0	1740.0

Manhours include unloading, job hauling up to 2000 feet, rigging, picking, setting, aligning, and checking out of conveyor and all components.

Manhours do not include installation of electrical circuits or connections.

CONVEYOR RECIPROCATING

INSTALLATION MANHOURS

Diameter Inches	MANHOURS	
	Steel	Stainless Steel
12	224.0	416.0
14	256.0	432.0
16	288.0	448.0
18	320.0	480.0
20	352.0	496.0
30	416.0	608.0
40	512.0	736.0
46	544.0	816.0

Manhours include unloading, job hauling up to 2000 feet, rigging, picking, setting, aligning, and checking out reciprocating conveyor.

Manhours exclude installation of electrical circuits and connections.

CONVEYOR SCROLL

INSTALLATION MANHOURS

Diameter Inches	MANHOURS	
	Bird Steel	Bird Stainless Steel
16	256.0	—
18	288.0	544.0
20	352.0	572.0
30	544.0	816.0
40	704.0	1040.0
50	880.0	1280.0
54	944.0	1360.0

Manhours include unloading, job hauling up to 2000 feet, rigging, picking, setting, aligning, and checking out of complete scroll.

Manhours exclude installation of electrical circuits and connections.

CONVEYORS–SPACED, BUCKET ELEVATOR

INSTALLATION MANHOURS

Length Feet	MANHOURS		
	(1)	(2)	(3)
25	99.2	132.0	208.0
30	108.8	176.0	224.0
40	126.4	208.0	256.0
50	142.4	232.0	280.0
60	175.2	256.0	304.0
70	176.0	272.0	336.0
80	192.0	280.0	368.0
90	208.0	288.0	392.0
100	224.0	296.0	424.0

(1) 6 x 4 x 4-1/2 inches;   14 tons per hour (100 pounds per cubic foot).

(1) 12 x 7 x 7-1/2 inches;   84 tons per hour.

(3) 16 x 7 x 7-1/2 inches; 150 tons per hour.

Manhours include unloading, job hauling up to 2000 feet, rigging, picking, setting, aligning, and checking out of conveyor and all components.

Manhours do not include installation of electrical circuits and connections.

Installation of conveyor drive, headshaft, tailshaft, chain or belt, buckets and casings such as head section, tail section, standard sections, and filler section are included in the manhours.

CONVEYORS-CONTINUOUS, BUCKET ELEVATOR

INSTALLATION MANHOURS

Length Feet	MANHOURS		
	(1)	(2)	(3)
25	113.6	176.0	240.0
30	126.4	188.0	256.0
40	165.6	224.0	272.0
50	176.0	248.0	296.0
60	200.0	264.0	336.0
70	216.0	272.0	368.0
80	232.0	280.0	384.0
90	256.0	288.0	408.0
100	272.0	296.0	440.0

(1) 8 x 5-1/2 x 7-3/4 inches, 35 tons per hour.

(2) 14 x 7 x 11-3/4 inches, 80 tons per hour.

(3) 16 x 8 x 11-3/4 inches, 115 tons per hour.

Manhours include unloading, job hauling up to 2000 feet, rigging, picking, setting, aligning, and checking out of conveyor and all components.

Manhours do not include installation of electric circuits and connections.

Installation of conveyor drive, headshaft, tailshaft, chain or belt, buckets and casings such as head section, tail section, standard sections, and filler section are included in the above manhours.

TUBE-FLO CONVEYORS

MANHOURS EACH

Item	Quantity	Installation Manhours Each		
		1	2	3
Drive Box	1 ea.	7.2	7.6	7.9
Discharge Gate	1 ea.	3.6	3.8	4.0
Inlet	1 ea.	3.6	3.8	4.0
Inspection Hatch	1 ea.	2.4	2.5	2.6
Compression Coupling	1 ea.	1.2	1.3	1.4
Chain and Flight	1 l.f.	0.2	0.2	0.3

1. Standard
2. Urethane
3. Cast iron

Manhours include unloading, handling, job hauling up to 2000 feet, rigging, picking, setting, aligning, and checking items. If the components have not been preassembled, additional field labor for fitup may be required. Additional labor may increase assembly manhours by 20 to 30%.

Manhours do not include pipe hangers and supports, electrical motor, electrical circuit, and electrical connections.

CRYSTALLIZERS–BATCH VACUUM

INSTALLATION MANHOURS

Working Capacity Gallons	MANHOURS		
	Steel	Rubber-Lined Steel	Stainless Steel
650	392.0	499.2	904.0
700	393.6	500.8	905.6
800	395.2	502.4	907.2
900	396.8	504.0	908.8
1000	398.4	512.0	912.0
2000	400.0	544.0	928.0
3000	424.0	576.0	960.0
4000	440.0	608.0	1024.0
5000	456.0	640.0	1104.0
6000	464.0	688.0	1184.0
7000	496.0	736.0	1280.0
8000	528.0	768.0	1424.0

Manhours include unloading, handling, job hauling up to 2000 feet, rigging, picking, setting, aligning, hooking up, and checking out of units as outlined.

Manhours exclude installation of supports and electrical power source.

CRYSTALLIZERS-MECHANICAL

INSTALLATION MANHOURS

Length Feet	MANHOURS	
	Steel & Cast Iron	Stainless Steel
20	140.8	280.0
30	192.0	336.0
40	256.0	416.0
50	280.0	496.0
60	316.0	576.0
70	344.0	632.0
80	400.0	720.0
90	424.0	784.0
100	456.0	816.0
200	768.0	1408.0
300	976.0	1920.0
400	1264.0	2480.0
500	1656.0	2800.0
600	1760.0	—
700	2000.0	—
800	2320.0	—
900	2560.0	—
1000	2720.0	—

Manhours include unloading, handling, job hauling up to 2000 feet, rigging, picking, setting, aligning, hooking up, and checking out of units as listed.

Manhours exclude installation of supports and electrical power source.

DOW THERM UNITS

INSTALLATION MANHOURS

Million BTU Per Hour Duty	Manhours Each
0.175	512.0
0.200	544.0
0.300	592.0
0.400	608.0
0.500	624.0
0.600	640.0
0.700	656.0
0.800	672.0
0.900	680.0
1.000	688.0
2.000	768.0
3.000	800.0
3.400	816.0

Manhours include unloading, handling, job hauling up to 2000 feet, rigging, picking, setting, aligning, hooking up, and checking out of units as listed.

Manhours exclude installation of piping, electrical circuits and their connections.

DRY MATERIAL BLENDERS

MANHOURS REQUIRED EACH

Motor Horsepower	Rotary Auger Type		Rotary Drum Type	
	Weight Pounds	Manhours Required	Weight Pounds	Manhours Required
1/2	400	24	–	–
3/4	600	24	–	–
1-1/2	850	24	–	–
2	975	24	2,000	24
5	–	–	3,100	40
5	–	–	3,600	40
10	–	–	6,500	48
15	–	–	9,200	48
20	–	–	12,500	72
25	–	–	15,000	72
40	–	–	28,000	84
75	–	–	36,000	84

Manhours include unloading, handling, job hauling up to 2000 feet, rigging, picking, setting, aligning, and checking out of item as outlined.

Manhours exclude installation of electrical circuits and connections.

Rotary auger type units are for mixing two or more dry materials such as powders, pellets, chunks, and fibers.

Rotary drum type units are for blending dry or semi-dry materials or combinations of both into a homogeneous whole.

VIBRATING PACKERS

MANHOURS REQUIRED EACH

Item Description	Motor Horsepower	Weight of Packer Pounds	Packing Weight Pounds	Manhours Required
Packers For Bag Packing				
Bag Width—14″	1/2	240	15/75	8
Bag Width—24″	1	720	50/150	10
Bag Width—30″	1	750	50/150	10
Bag Width—36″	1	760	50/150	10
Packers For Rigid Containers				
Container Size—11-5/8″ sq./10-1/2″ dia.	1/2	220	15/75	8
Container Size—19″ - 21″ - 25″ dia.	1	760	100/1000	10

Manhours include unloading, handling, job hauling up to 2000 feet, rigging, picking, setting, aligning, and checking out of item as outlined.

Manhours exclude installation of electrical circuits and connections.

Bag packers are for use in conjunction with belt conveying systems and automatic scale and closing equipment and is designed for packing burlap, cotton, or paper bags.

Rigid container packers are so designed for packing kegs, barrels, drums, or other type rigid containers.

DRY MATERIAL FEEDERS

MANHOURS REQUIRED EACH

Item Description	Approximate Weight Pounds	Manhours Required
Electric Vibrating Feeders		
Small Feeder with Vibrating Hopper	75	4
Small Feeder with 30" Long Open Pan Deck	235	6
Medium Feeder with 42" Long Open Pan Deck	1,000	10
Heavy Duty Feeder with 60" Long Open Pan Deck	4,000	24
Extra Heavy Duty Feeder with 72" Long Open Pan Deck	7,600	32
Gravimetric Feeders		
Small	750	18
Medium	1,200	24
Large	1,350	24
Volumetric Feeders		
Mechanical Variable Speed Drive	350	8
DC Drive System & Solid State SCR Controller	350	10
Wing-Type Feeders		
Rachet Drive Type	300/600	8
Rachet Drive Type	650/1,200	10
Micro-Drive Type	400/700	10
Micro-Drive Type	750/1,250	12
Conveyor Type Feeders		
Feeder	300/650	8
Sanitary Type Feeders		
Feeder	400/700	10
Feeder	750/1,250	12
Dry Polymer Feeder		
Feeder	90/250	14

Manhours include unloading, handling, job hauling up to 2000 feet, rigging, picking, setting, aligning, and checking out of feeder.

Manhours exclude installation of electric circuits and connections.

DRYERS—ATMOSPHERIC DRUM

INSTALLATION MANHOURS

Approximate		Motor	MANHOURS	
Peripheral Square Feet Area	Weight Pounds	Horsepower Range	Single Drum	Double Drum
12.5	5,000	2-5	176.0	–
22.0	10,000	5-10	202.0	–
25.0	8,500	2-5	–	288.0
33.0	11,250	5-10	220.0	–
37.6	9,200	2-5	–	336.0
44.0	12,500	5-10	246.0	–
50.2	10,000	2-5	–	360.0
55.0	13,750	5-15	264.0	–
66.0	15,000	5-15	290.0	–
72.6	16,800	2-10	–	432.0
100.0	18,400	3-15	–	480.0
126.0	19,600	3-15	–	528.0
139.0	20,500	3-15	–	552.0
157.0	34,000	15-100	400.0	–
165.0	32,500	5-20	–	660.0
167.0	22,300	3-20	–	600.0
183.0	34,000	5-20	–	660.0
188.0	36,000	20-150	460.0	–
220.0	37,000	5-20	–	720.0
251.0	40,000	15-50	500.0	–
377.0	60,000	10-30	–	756.0

Manhours include unloading, handling, job hauling up to 2000 feet, assembling as may be required, rigging, picking, setting, aligning, and checking out of dryer as outlined.

Manhours exclude installation of electrical circuits and connections.

Single atmospheric drum manhours are for installing unit consisting of steel base mounted, cast iron chrome plated drum with stainless steel applicator rolls, drive, and motor.

Double atmospheric drum manhours are for installing unit consisting of steel base mounted twin cast iron chrome plated drums, vapor hood, stainless steel side and cross conveyors, conveyor drive, elevator and flaker, drive, and motor.

DRYERS—VACUUM DRUM

INSTALLATION MANHOURS

Approximate		Motor	Manhours	
Peripheral Square Feet Area	Weight Pounds	Horsepower Range	Single Drum	Double Drum
10.4	7,700	2-5	220.0	–
25.0	17,000	2-5	–	229.0
42.0	31,000	5-10	525.0	–
50.0	22,500	2-5	–	246.0
100.0	41,000	3-15	–	950.0
188.0	75,000	10-30	952.0	–
220.0	87,500	5-20	–	1,140.0

Manhours include unloading, handling, job hauling up to 2000 feet, assembling as may be required, rigging, picking, setting, aligning, and checking out of dryer as outlined.

Manhours exclude installation of electrical circuits and connections.

Vacuum drum manhours are for installing single or double drum units consisting of drums, feed devices, product removal knives, dry material conveyors enclosed in an air-tight casing, and installation of conveyor drive and drum drive and motor.

DRYERS—TWIN DRUM

INSTALLATION MANHOURS

| Approximate | | Motor | |
Peripheral Square Feet Area	Weight in Pounds	Horsepower Range	Manhours
25.0	8,500	2 - 5	345.0
37.6	9,200	2 - 5	393.0
72.6	16,800	2 - 10	505.0
100.0	18,400	3 - 15	561.0
126.0	19,600	3 - 15	618.0
139.0	20,500	3 - 15	646.0
165.0	32,500	5 - 20	772.0
167.5	22,300	3 - 20	703.0
183.5	34,000	5 - 20	772.0
220.0	37,000	5 - 20	842.0
377.0	60,000	10 - 30	885.0

Manhours include unloading, handling, job hauling up to 2000 feet, rigging, picking, setting, aligning, and checking out of dryer as outlined.

Manhours exclude installation of electrical circuits and connections.

Twin drum dryers differ from double drum units in that the drums rotate away from rather than toward the pinch, knives are mounted in the lower outer quadrant, and feed can be applied from above or below the drum.

Manhours are for installing units consisting of base, drums, side and cross conveyors, conveyor drive, and drum drive and motor.

If vapor hood is to be installed, add 40 manhours for this operation.

DRYERS-TRAY, ATMOSPHERIC

INSTALLATION MANHOURS

Top Tray Square Feet Area	MANHOURS	
	Steel	Stainless Steel
30	48.0	81.6
40	51.2	91.2
50	56.0	97.6
60	57.6	104.0
70	59.2	110.4
80	60.8	113.6
90	62.4	120.0
100	63.2	124.0
200	72.0	155.2
250	—	176.0

Manhours include unloading, handling, job hauling up to 2000 feet, rigging, picking, setting, aligning, and checking out of dryer as outlined.

Manhours exclude installation of electrical circuits and connections.

DRYERS-TRAY, VACUUM

INSTALLATION MANHOURS

Top Tray Area Square Feet	MANHOURS	
	Steel	Stainless Steel
40	81.6	192.0
50	88.0	216.0
60	92.8	248.0
70	97.6	272.0
80	102.4	288.0
90	108.8	304.0
100	110.4	316.0
200	136.0	448.0

Manhours include unloading, handling, job hauling up to 2000 feet, rigging, picking, setting, aligning, and checking out of dryer as outlined.

Manhours exclude installation of electrical circuits and connections.

DRYERS-ROTARY

INSTALLATION MANHOURS

Peripheral Square Feet Area	MANHOURS			
	Hot Air	Flue Gas Direct	Flue Gas Indirect	Vacuum
100	153.6	208.0	256.0	624.0
200	256.0	304.0	384.0	778.0
300	304.0	432.0	504.0	880.0
400	368.0	504.0	632.0	968.0
500	432.0	600.0	736.0	1040.0
600	472.0	656.0	816.0	—
700	512.0	752.0	928.0	—
800	576.0	808.0	992.0	—
900	616.0	912.0	1112.0	—
1000	648.0	952.0	1168.0	—
2000	976.0	1520.0	1840.0	—

Manhours include unloading, handling, job hauling up to 2000 feet, rigging, picking, setting, aligning, and checking out of dryer as outlined.

Manhours exclude installation of electrical circuits and connections.

DRYERS-SPRAY

INSTALLATION MANHOURS

Water Pounds Per Hour Evaporative Capacity	MANHOURS		
	10 Feet Diameter	14 Feet Diameter	18 Feet Diameter
700	800.0	–	–
800	800.0	–	–
900	808.0	–	–
1000	816.0	–	–
2000	928.0	1184.0	–
3000	1008.0	1232.0	1552.0
4000	–	1272.0	1592.0
5000	–	1312.0	1608.0
6000	–	1368.0	1680.0
7000	–	–	1760.0
8000	–	–	1840.0
9000	–	–	1856.0

Manhours include unloading, handling, job hauling up to 2000 feet, rigging, picking, setting, aligning, and checking out of dryer as outlined.

Manhours exclude installation of electrical circuits and connections.

COOLING DRUM FLAKERS

INSTALLATION MANHOURS

Approximate Dimensions			Drum Surface	Motor Horsepower	Approximate Weight	
Length	Width	Height	Square Feet	Range	Pounds	Manhours
5'8"	3'6"	4'0"	10.4	1/2 - 2	2,600	48.0
6'0"	3'6"	4'0"	12.5	3/4 - 2	2,700	51.0
6'6"	3'6"	4'0"	15.7	3/4 - 2	2,800	55.0
7'0"	3'6"	4'0"	18.8	1 - 3	3,000	57.0
8'0"	3'6"	4'0"	25.0	1 - 3	3,300	62.0
9'0"	3'6"	4'0"	31.4	1 - 3	3,600	64.0
5'8"	5'0"	6'0"	29.0	1 - 3	6,900	70.0
6'8"	5'0"	6'0"	42.0	1 - 3	7,500	77.0
7'4"	5'0"	6'0"	50.0	1 - 5	8,000	80.0
8'7"	5'0"	6'0"	63.0	1-1/2 - 5	8,700	90.0
9'7"	5'0"	6'0"	75.0	1-1/2 - 5	9,300	93.0
10'7"	5'0"	6'0"	88.0	1-1/2 - 5	10,000	96.0
11'7"	5'0"	6'0"	100.0	2 - 7-1/2	10,600	99.0
12'7"	5'0"	6'0"	112.0	2 - 7-1/2	11,300	106.0
13'7"	5'0"	6'0"	125.0	2 - 10	12,000	112.0
15'7"	5'0"	6'0"	150.0	3 - 10	14,000	122.0
9'7"	6'0"	7'0"	94.0	2 - 7-1/2	17,500	134.0
11'7"	6'0"	7'0"	125.0	2 - 10	21,000	147.0
13'7"	6'0"	7'0"	157.0	3 - 10	25,000	160.0
15'7"	6'0"	7'0"	188.0	3 - 15	29,000	173.0

Manhours include unloading, handling, job hauling up to 2000 feet, rigging, picking, setting, aligning and checking out of cooling drum flakers as outlined.

Manhours exclude installation of electrical circuits and connections.

Cooling drum flaker units consist of frame, drum, liquid spray drum cooling system, knife and holder, and drive.

The listed approximate weight does not include the weight of the drive assembly.

DUST COLLECTORS

Cyclone and Multiple Cyclone

INSTALLATION MANHOURS

Capacity Cubic Feet Per Minute	MANHOURS	
	Cyclones	Multiple Cyclones
900	30.4	46.4
1,000	33.6	48.0
2,000	36.8	54.4
3,000	41.6	57.6
4,000	43.2	59.2
5,000	44.8	60.8
6,000	46.4	62.4
7,000	47.2	64.0
8,000	48.0	65.6
9,000	49.6	67.2
10,000	50.4	68.0
20,000	57.6	72.8

Manhours include unloading, handling, job hauling up to 2000 feet, rigging, picking, assembling, setting, aligning, and checking out of collectors as listed.

Manhours exclude installation of electrical circuits and connections.

Installation of the cyclone, dust hopper, scroll outlet, weather cap, and support stand have been included where required in the manhours.

DUST COLLECTORS
Centrifugal Precipitator and Automatic Cloth Filter Types

INSTALLATION MANHOURS

Capacity Cubic Feet Per Minute	MANHOURS		
	Washers	Centrifugal Precipitators	Automatic Cloth Filters
600	56.0	64.0	75.2
700	59.2	68.8	80.0
800	62.4	72.0	84.8
900	65.6	75.2	89.6
1,000	68.8	78.4	94.4
2,000	94.4	104.0	128.0
3,000	112.0	126.4	153.6
4,000	126.4	162.4	192.0
5,000	139.2	156.8	216.0
6,000	150.4	172.0	240.0
7,000	160.0	192.0	256.0
8,000	176.0	208.0	264.0
9,000	194.0	224.0	272.0
10,000	208.0	240.0	280.0
20,000	272.0	288.0	352.0

Manhours include unloading, handling, job hauling up to 2000 feet, picking, assembling, setting, aligning, and checking out of collectors as listed.

Manhours exclude installation of electrical circuits and connections.

Manhours for the above wet collectors include the complete installation of the collector excluding the installation of the spray nozzles and piping.

Automatic cloth filter manhours include the installation of the baghouse with bags, continuous hopper, structural support ladder, and walkway.

DUST COLLECTORS–ELECTRICAL PRECIPITATORS

INSTALLATION MANHOURS

Capacity Cubic Feet Per Minute	MANHOURS	
	Low Voltage	High Voltage
600	22.4	–
700	27.2	–
800	28.8	–
900	30.4	–
1,000	33.6	–
2,000	62.4	–
3,000	91.2	–
4,000	113.6	–
5,000	142.4	–
6,000	168.0	–
7,000	200.0	448.0
8,000	240.0	464.0
9,000	264.0	472.0
10,000	280.0	488.0
20,000	496.0	608.0

Manhours include unloading, handling, job hauling up to 2000 feet, rigging, picking, assembling, setting, aligning, and checking out of collectors as listed.

Manhours exclude installation of electrical circuits and connections.

DUST COLLECTOR—FEEDER VALVES

INSTALLATION MANHOURS

Valve Type	Size Inches	Manhours Required
Rotary Feeder	5	5.0
Rotary Feeder	6	5.2
Rotary Feeder	7	7.0
Rotary Feeder	8	7.0
Rotary Feeder	10	9.0
Rotary Feeder	12	11.1
Counter Weighted, Dual Trickle, Gravity Operated	12	13.1
Counter Weighted, Dual Trickle, Motor Operated	12	18.5
Counter Weighted, Dual Trickle, Cylinder Op. with Timer	12	20.5

Manhours include unloading, handling, job hauling up to 2000 feet, positioning, and connecting of feeder values for use with the collectors as outlined.

Manhours do not include installation of dust collectors, piping, or electrical circuits and connections.

FILTER CARTRIDGE DUST COLLECTORS

Combine Cartridge-Type Filters With Pulse Jet
Continuous-Duty Dust Collectors

MANHOURS EACH

Maximum Size Inches			Cartridge Filter Area	Cartridges	Shipping Weight	Manhours
Height	Depth	Width	Square Feet	Each	Pounds	Each
108⅛	36¾	36¾	162	3	426	58
116½	28½	28½	486	9	375	63
150⅜	60	59	573	3	1,350	77
141½	68	52	1,150	6	3,200	96
141½	94	52	2,300	12	3,500	104
173⅜	94	78	3,060	16	4,000	112
180⅞	94	104	4,600	24	4,500	115
180⅞	94	130	6,120	32	5,000	127

Manhours include unloading, handling, job hauling up to 2000 feet, picking, assembling, setting, aligning, and checking collectors as listed.

Manhours exclude installation of related blowers, electric motors, and their connections.

FILTER CARTRIDGE DUST COLLECTORS

With Quick-Change Filter Elements

MANHOURS EACH

Maximum Size Inches			Cartridge Filter Area	Cartridges	Shipping Weight	Manhours
Height	Depth	Width	Square Feet	Each	Pounds	Each
124⅞	85¼	40	1,808	8	1,850	58
124⅞	85¼	58³⁄₁₆	2,712	12	2,430	60
124⅞	85¼	80	3,616	16	3,020	77
143½	85¼	80	5,424	24	3,730	81
162⅛	85¼	80	7,232	32	4,470	96
143½	85¼	120	8,136	36	5,230	106
162⅛	85¼	120	10,848	48	6,290	115
143½	85¼	200	13,560	60	8,860	125
143½	85¼	240	16,272	72	9,660	131
162⅛	85¼	200	18,080	80	10,170	144

Manhours include unloading, handling, job hauling up to 2000 feet, picking, assembling, setting, aligning and checking collectors as listed.

Manhours exclude installation of related blowers, electrical motors, and their connections.

FILTER CARTRIDGE SYSTEM DUST COLLECTORS

MANHOURS EACH

Maximum Size Inches			Cartridge Filter Area	Cartridges	Shipping Weight	Manhours
Height	Depth	Width	Square Feet	Each	Pounds	Each
102¼	96	161⁷⁄₁₆	8,136	36	6,350	96
102¼	96	196¹⁄₁₆	10,848	48	7,600	101
102¼	96	231⁷⁄₁₆	13,560	60	8,850	125
102¼	96	266⁷⁄₁₆	16,272	72	10,600	144
102¼	96	301⁷⁄₁₆	18,984	84	11,850	154

Note: System removes airborne particles as fine as 0.5 micron, welding fumes, cement, coal dust, fly ash, chemical dusts, and other industrial contaminants. Units are factory assembled (up to seven modules) and shipped intact.

Manhours include unloading, handling, job hauling up to 2000 feet, picking, assembling, setting, aligning, and checking collectors as listed.

Manhours exclude installation of related blowers, electrical motors, and their connection.

EJECTORS–STEAM JET

Single-Stage Noncondensing
50, 76, and 102 Millimeters Mercury

INSTALLATION MANHOURS

Capacity Pounds Air Per Hour 100 psi Steam 70° F. Air	MANHOURS		
	Pressure Millimeters of Mercury 50	Pressure Millimeters of Mercury 76	Pressure Millimeters of Mercury 102
10	8.0	—	—
12	—	8.0	—
20	10.4	9.2	8.0
30	12.0	10.0	8.8
40	13.6	10.8	9.2
50	14.8	11.6	10.0
60	16.0	11.8	10.4
70	—	12.4	10.8
80	—	12.8	11.2
90	—	13.2	11.6
100	—	13.6	12.0
200	—	16.0	14.4
300	—	—	16.0

Manhours include unloading, handling, job hauling up to 2000 feet, rigging, picking, setting, aligning, and checking out of items as outlined.

Manhours exclude installation of piping and instrumentation.

EJECTORS–STEAM JET

Single-Stage Noncondensing
152, 203, and 304 Millimeters Mercury

INSTALLATION MANHOURS

Capacity Pounds Air Per Hour 100 psi Steam 70° F. Air	MANHOURS		
	Pressure Millimeters of Mercury 152	Pressure Millimeters of Mercury 203	Pressure Millimeters of Mercury 304
30	8.0	–	–
40	8.4	–	–
50	8.8	8.0	–
60	9.2	8.4	–
70	9.6	8.6	–
80	10.0	9.0	8.0
90	10.4	9.2	8.4
100	10.8	9.6	8.6
200	12.4	11.2	10.4
300	13.6	12.8	11.2
400	14.4	13.6	12.4
500	15.2	14.0	12.8
600	16.0	14.8	13.6
700	–	15.6	14.0
800	–	16.0	14.8
900	–	–	15.6
1000	–	–	16.0

Manhours include unloading, handling, job hauling up to 2000 feet, rigging, picking, setting, aligning, and checking out of items as outlined.

Manhours exclude installation of piping and instrumentation.

EJECTORS–STEAM JET

Two-Stage, Barometric Type Intercondenser

INSTALLATION MANHOURS

Capacity Pounds Air Per Hour 100 psi Steam 70° F. Air	MANHOURS			
	Pressure Millimeters of Mercury 100	Pressure Millimeters of Mercury 50	Pressure Millimeters of Mercury 25	Pressure Millimeters of Mercury 10
18	–	16.0	20.0	27.2
20	16.0	18.4	22.4	28.8
30	20.0	21.6	25.0	31.2
40	22.4	24.0	26.4	36.8
50	24.0	25.6	28.0	40.0
60	25.0	27.2	29.6	44.0
70	26.0	28.8	32.8	–
80	26.8	30.4	34.4	–
90	27.6	32.0	36.0	–
100	28.8	32.8	38.4	–
125	30.4	–	–	–

Manhours include unloading, handling, job hauling up to 2000 feet, rigging, picking, setting, aligning, and checking out of items as outlined.

Manhours exclude installation of piping and instrumentation.

EXTRACTORS-CONTINUOUS CENTRIFUGAL

INSTALLATION MANHOURS

Capacity Gallons Per Minute	Manhours Each
4	192.0
5	208.0
6	224.0
7	240.0
8	256.0
9	272.0
10	288.0
20	328.0
30	432.0
40	496.0

Manhours include unloading, handling, job hauling up to 2000 feet, rigging, picking, setting, aligni and checking out of items as outlined.

Manhours exclude installation of supports and piping.

HEAVY-GAUGE CENTRIFUGAL FANS

MANHOURS REQUIRED EACH

Diameter of Wheel Inches	Maximum CFM Range	Approximate Weight Pounds	Installation Manhours
12-1/4	2,100	140	4.40
13-1/2	2,700	165	4.50
15	2,900	185	5.70
16-1/2	3,500	195	6.00
18-1/4	5,000	235	6.40
20	6,500	300	6.50
22-1/4	8,000	330	7.30
24-1/2	9,100	425	7.40
27-1/2	12,000	525	7.55
30	14,750	610	8.00
33	18,000	775	8.40
36-1/2	25,000	940	9.60
40-1/4	30,000	1,575	12.00
44-1/2	37,000	1,870	13.80
49	45,000	2,225	16.80
54-1/4	55,000	2,750	22.60
60	67,000	3,050	30.00
66	81,000	3,900	43.40
73	99,000	5,625	56.50
80-3/4	122,000	6,850	70.70
89	148,000	8,800	82.80

Manhours include receiving and unloading at job site, moving within 50 feet of erection site, setting and aligning at floor level, and adjusting bearings.

Manhours do not include installation of inlet vane control, out dampers, motor and drives. See respective tables for these time frames.

FAN MOTORS & V-BELT DRIVES

MANHOURS REQUIRED FOR ITEMS LISTED

Motor Horsepower	Motor Weight Pounds	INSTALLATION MANHOURS	
		Fan Motors	V-Belt Drives
3	85	3.56	1.00
5	95	4.28	1.50
7-1/2	125	4.99	2.00
10	175	6.41	2.20
15	230	7.84	2.40
20	350	9.98	2.60
25	465	11.40	3.00
30	510	14.25	4.50
40	600	18.53	4.80
50	670	21.38	6.00
60	745	24.23	7.40
75	820	28.50	9.00
100	875	34.20	–

Fan motor manhours include mounting of motor on base and adjusting drive alignment.

V-belt manhours include installation of belt and alignment of wheels.

All manhours include receiving and off-loading at job site, uncrating, and moving within 50 feet of erection site.

Manhours do not include electrical hook-up. See respective table for this time requirement.

BLOWERS-ROTARY

10 to 15 PSI

INSTALLATION MANHOURS

Capacity Cubic Feet Per Minute	Manhours Each
100	14.4
200	19.4
300	24.3
400	27.5
500	29.2
600	30.8
700	31.5
800	34.0
900	37.4
1,000	47.9
2,000	50.2
3,000	52.6
4,000	66.1
5,000	72.9
6,000	77.8
7,000	82.6
8,000	87.5
9,000	94.0
10,000	97.2

Manhours include handling, hauling, setting, aligning, hook-up and testing of blower and components.

Manhours do not include installation of electrical circuit to blowers.

BLOWERS–CENTRIFUGAL TURBO

INSTALLATION MANHOURS

Capacity Cubic Feet Per Minute	MANHOURS		
	0.5 to 2 psi	7 to 10 psi	20 to 30 psi
100	14.6	—	—
200	24.3	—	—
300	32.6	—	—
400	38.9	—	—
500	45.4	—	—
600	56.0	—	—
700	63.0	—	—
800	68.0	—	—
900	71.8	—	—
1,000	77.4	112.5	—
2,000	122.4	178.2	—
3,000	160.2	216.0	774.0
4,000	198.0	252.0	810.0
5,000	—	270.0	846.0
6,000	—	279.0	882.0
7,000	—	292.5	900.0
8,000	—	306.0	918.0
9,000	—	315.5	936.0
10,000	—	333.0	963.0
20,000	—	432.0	1062.0
30,000	—	522.0	1116.0
40,000	—	576.0	—

Manhours include handling, hauling, setting, aligning, hook-up and testing of blower and components.

Manhours do not include installation of electrical circuits.

FILTERS—PRESSURE TYPE

MANHOURS REQUIRED EACH

Filter Diameter Feet	Overall Height Inches	Approximate Weight Pounds	Manhours
1	71	500	23.4
2	71	900	27.0
3	83	1,600	29.7
4	89	2,500	33.8
5	104	4,000	39.0
6	109	8,000	50.4
7	114	10,500	57.6
8	120	14,000	70.2
9	123	19,000	80.6
10	127	23,500	88.4

Manhours include unloading, handling, job hauling up to 2000 feet, rigging, picking, setting, aligning, and checking out of filter components and pipe, valves, and fittings from filter inlet to outlet.

Manhours exclude installation of pipe lines from or to filter inlet and outlet.

Pressure filters are of noncode steel shell contruction with interior filtering materials and are primarily used for the pretreatment of raw water for domestic or industrial use. They are designed to remove dirt, rust, oil, color, taste, suspended solids and turbidity from process water and to move residual, biological solids, chemical precipitates, and other suspended solids from waste water.

FILTERS—OIL MIST COLLECTORS

MANHOURS REQUIRED EACH

Capacity CFM	Approximate Wt. Pounds	Manhours
4,000	840	24.0
8,000	1,400	36.0
12,000	1,920	46.0
16,000	2,450	72.0
20,000	3,010	78.0
24,000	3,500	94.0
28,000	4,100	112.0
32,000	3,950	94.0

Manhours include unloading, handling, job hauling up to 2000 feet, rigging, picking, setting, aligning, and checking out of inlet and plenum section, outlet section, first stage filter section and second stage filter section.

Manhours exclude installation of other duct work.

Collectors are modular units for use of high efficiency separation of mists and fogs from air or process gases.

FILTERS–PLATE AND FRAME

Cast Iron, Yellow Pine, Aluminum

INSTALLATION MANHOURS

Filtering Area Square Feet	MANHOURS		
	Cast Iron	Yellow Pine	Aluminum
18	11.2	13.0	17.6
20	12.2	14.4	20.0
30	14.2	16.8	25.6
40	16.0	20.0	27.2
50	19.2	21.6	28.8
60	20.8	23.2	30.4
70	22.4	24.8	32.8
80	24.0	26.4	34.4
90	25.6	27.2	36.8
100	27.2	28.8	40.0
200	30.4	30.4	54.4
300	38.4	37.6	64.0
400	44.8	44.0	72.0
500	46.4	45.6	80.0
600	49.6	48.0	86.4
700	54.4	52.8	94.4
800	57.6	56.0	97.6
900	60.8	59.2	104.0
1000	62.4	60.8	110.4
1500	72.0	69.6	128.0

Manhours include unloading, handling, job hauling, assembling as necessary, rigging, picking, setting, aligning, and checking out of filter as outlined.

Manhours exclude installation of supporting structure.

FILTERS–PLATE AND FRAME

Lead, Bronze, Stainless Steel

INSTALLATION MANHOURS

Filtering Area Square Feet	MANHOURS		
	Lead	Bronze	Stainless Steel
18	25.6	33.6	63.2
20	29.2	40.0	70.4
30	30.4	44.8	80.8
40	35.2	50.4	94.4
50	38.4	57.6	102.4
60	41.6	60.8	111.2
70	44.8	64.0	119.2
80	46.4	67.2	127.2
90	49.6	70.4	132.8
100	51.2	73.6	140.8
200	69.6	96.0	192.0
300	81.6	113.6	224.0
400	92.8	129.6	256.0
500	102.4	144.0	288.0
600	108.4	158.4	304.0
700	118.4	168.0	320.0
800	126.4	184.0	336.0
900	129.6	192.0	368.0
1000	127.6	208.0	384.0
1500	160.0	240.0	448.0

Manhours include unloading, handling, job hauling up to 2000 feet, assembling as necessary, rigging, picking, setting, aligning, and checking out of filter as outlined.

Manhours exclude installation of supporting or other structures.

FILTERS–SPARKLER

INSTALLATION MANHOURS

Diameter Inches	MANHOURS		
	Carbon Steel	Stainless Steel	Hastelloy
14	19.2	38.4	84.8
15	22.4	46.4	102.4
16	25.6	51.2	120.0
17	28.8	59.2	142.4
18	30.4	62.4	158.4

Manhours include unloading, handling, job hauling up to 2000 feet, rigging, picking, setting, aligning, and checking out of filter as outlined.

Manhours exclude installation of supporting or other structures.

FILTERS-LEAF

INSTALLATION MANHOURS

Filtering Area Square Feet	MANHOURS	
	2-inch Spacing	4-inch Spacing
35	65.6	81.6
40	68.8	86.4
50	75.2	96.0
60	78.4	104.0
70	81.0	110.4
80	88.0	113.6
90	91.2	121.6
100	94.4	128.0
200	120.0	168.0
300	140.8	208.0
400	152.0	240.0
500	168.0	256.0
600	176.0	272.0
700	192.0	288.0
800	208.0	304.0
900	224.0	320.0
1000	240.0	328.0
1500	256.0	352.0

Manhours include unloading, handling, job hauling up to 2000 feet, assembling where necessary, rigging, picking, setting, aligning, and checking out of filter as outlined.

Manhours exclude installation of supports or other structures.

FILTERS–SEWAGE AND ROTARY

INSTALLATION MANHOURS

Filtering Area Square Feet	MANHOURS		
	Sewage	Rotary Drum	Rotary Disk
100	288.0	—	—
200	304.0	496.0	544.0
300	318.4	592.0	672.0
400	336.0	656.0	792.0
500	360.0	736.0	896.0
600	384.0	784.0	976.0
700	—	816.0	1104.0
800	—	880.0	1184.0
900	—	928.0	1280.0
1000	—	960.0	1360.0
1500	—	1040.0	1584.0

Manhours include unloading, handling, job hauling up to 2000 feet, assembling where necessary, rigging, picking, setting, aligning, and checking out of filters as outlined.

Manhours exclude installation of supports or other structures.

FILTERS—SCREEN VIBRATING

INSTALLATION MANHOURS

Screen Area Square Feet	MANHOURS		
	Single Deck	Double Deck	Triple Deck
4	16.0	19.2	22.4
5	16.8	16.8	23.2
6	17.6	20.8	24.0
7	18.4	21.6	24.8
8	19.2	22.4	25.6
9	20.0	23.2	26.4
10	20.8	24.0	27.2
20	23.2	26.4	29.6
30	24.0	28.8	33.6
40	25.6	30.4	36.8
50	28.0	33.6	41.6
60	32.0	38.4	46.4
70	40.0	44.0	52.8
80	48.0	51.2	60.8

Manhours include unloading, handling, job hauling up to 2000 feet, assembling as required, rigging, picking, setting, aligning, and checking out of filter as outlined.

Manhours exclude installation of supports or other items.

RUBBER-LINED FILTERS

MANHOURS EACH

Flow Rating		For Tanks Up To Gallons	Pump HP	Dimensions L × W × H Inches	Approximate Weight Pounds	Manhours Each
GPM	GPH					
10	600	400	1	21 × 18 × 46	200	3
20	1,200	800	1	26 × 23 × 40	250	3
35	2,000	1,400	1	26 × 23 × 48	300	4
50	3,000	2,000	2	38 × 32 × 52	600	5
75	4,500	3,000	2	38 × 32 × 52	650	5
100	6,000	4,000	3	38 × 40 × 52	700	6
150	9,000	6,000	5	48 × 40 × 52	900	8
20	1,200	800	¾	40 × 36 × 44	550	29
35	2,100	1,400	1	40 × 36 × 50	625	31
50	3,000	2,000	2	60 × 50 × 59	1,000	34
75	4,500	3,000	2	60 × 50 × 59	1,050	34
100	6,000	4,000	3	60 × 50 × 59	1,100	38
150	9,000	6,000	5	66 × 50 × 59	1,400	40

Note: All items above the intermediate line are mounted on casters. All items below the intermediate line in addition to the rubber-lined filter and pumps have adequate sized rubber-lined steel slurry tanks, rubber-lined Hills-McCanna valves and rubber-lined steel fittings.

Manhours include unloading, handling, job hauling up to 2000 feet, rigging, picking, setting, aligning, and checking filters.

Manhours exclude installation of main piping lines and electrical circuits.

FLOTATION MACHINES

INSTALLATION MANHOURS

Capacity Cubic Feet	Manhours
3	26.4
4	27.2
5	28.0
6	28.8
7	29.6
8	30.4
9	31.2
10	32.0
20	38.4
30	44.8
40	49.6
50	57.6
60	62.4
70	70.4
80	86.4
90	88.0
100	97.6

Manhours include unloading, handling, job hauling up to 2000 feet, rigging, picking, setting, aligning, and checking out of items as outlined.

Manhours exclude installation of supports and piping.

GAS HOLDERS

INSTALLATION MANHOURS

Capacity Cubic Feet	Manhours
1,000	80.0
2,000	112.0
3,000	131.2
4,000	152.0
5,000	168.0
6,000	192.0
7,000	216.0
8,000	240.0
9,000	256.0
10,000	264.0
20,000	320.0
30,000	368.0
40,000	448.0
50,000	480.0
60,000	512.0
70,000	576.0
80,000	608.0
90,000	632.0
100,000	656.0
200,000	896.0
300,000	1088.0
400,000	1248.0
500,000	1344.0
600,000	1640.0
700,000	1584.0
800,000	1632.0
900,000	1696.0
1,000,000	1760.0

Manhours include unloading, handling, job hauling up to 2000 feet, rigging, picking, setting, aligning, and checking out of units as outlined.

Manhours exclude installation of supports and piping.

GAS SCRUBBER—HIGH ENERGY VENTURI

INSTALLATION MANHOURS

Approximate Weight Pounds	Gas Range Capacity Maximum	ACFM Minimum	Manhours Each
35	360	225	10
50	550	340	10
80	860	535	10
110	1,235	770	10
170	1,935	1,205	19
245	2,785	1,735	19
420	3,790	2,360	19
595	5,375	3,345	29
855	7,740	4,820	29
1,165	10,535	6,560	29
1,520	13,760	8,570	29
2,615	17,415	10,850	38
3,230	21,500	13,395	38
3,905	26,015	16,205	38
4,650	30,960	19,290	38
5,455	36,330	22,635	58
6,325	42,140	26,255	58
8,260	55,035	34,290	77
10,455	69,655	43,400	77
12,910	85,895	53,580	115
21,030	104,055	64,835	115
25,025	123,835	77,160	154
29,370	145,335	90,555	154
36,540	180,810	112,660	173

Manhours include unloading, handling, job hauling up to 2000 feet, picking, assembling, setting, aligning, and checking collectors as listed.

Manhours exclude tie-ins to other process systems.

STEAM TURBINE GENERATOR UNITS
Condensing, Noncondensing, Condensing Automatic Extraction, and Noncondensing Automatic Extraction

General Notes

Unload and Handling – Units include unloading at erection site or building, jacking, cribbing, skidding, and moving to erection location.

Plates – Units include installation, grouting where necessary and aligning of sub-soles and bedplates.

Lift Set and Align – Units include setting, aligning and hook-up of all components for low pressure turbine, I.P.-L.P. Turbine, high pressure turbine, generator and exciter.

Piping – Units include installation of oil, steam and generator piping and fittings and related instrumentation.

Oil Flush and Test – Units include oil flush, generator test, bearing inspection, balance test and close up for start-up.

Electrical – Units include all electrical and related instrumentation installation.

Insulation and Lagging – Units include installation of all insulation and lagging.

Start – Units include start-up and final test.

Final – All units are for installation of all components for the described items necessary for the complete operation of the turbine generator. Piping is included from the main block valves through the turbine generator only.

STEAM TURBINE GENERATOR UNITS
Weight Tables

APPROXIMATE WEIGHT OF UNITS IN THOUSANDS OF POUNDS

Item	Listed KW Rating					
	2000	2500	3000	4000	5000	6000
Unit net weight	90	90	93	109	118	129
Unit shipping weight	100	100	102	120	130	142
Heaviest piece to erect	45	45	45	50	55	60

Item	Listed KW Rating				
	7500	10,000	12,650	15,625	16,500
Unit net weight	134	187	206	216	233
Unit shipping weight	147	206	226	237	256
Heaviest piece to erect	65	100	105	119	140

STEAM TURBINE GENERATOR UNITS

Condensing, Non-condensing, Condensing Automatic Extraction, and Non-condensing Automatic Extraction

INSTALLATION MANHOURS

Item of Work	Manhours Per Listed KW Rating		
	2000	2500	3000
Unloading and handling	200.	200.	200.
Plates	600.	600.	700.
Lift, set and align	6,200.	6,200.	6,630.
Piping	5,100.	5,100.	6,100.
Oil flush and test	1,550.	1,550.	1,800.
Electrical	1,500.	1,500.	1,750.
Insulation and lagging	1,500.	1,500.	2,000.
Start-up	500.	500.	550.
Total	17,150.	17,150.	19,730.

Item or Work	Manhours Per Listed KW Rating		
	4000	5000	6000
Unload and handling	225.	250.	300.
Plates	750.	825.	825.
Lift, set and align	6,850.	7,350.	7,350.
Piping	6,500.	7,300.	7,300.
Oil flush and test	2,000.	2,175.	2,175.
Electrical	1,800.	1,850.	1,850.
Insulation and lagging	2,150.	2,125.	2,125.
Start-up	600.	650.	650.
Total	20,875.	22,525.	22,575.

See General Notes for explanation.

STEAM TURBINE GENERATOR UNITS

Condensing, Non-condensing, Condensing Automatic Extraction, and Non-condensing Automatic Extraction

INSTALLATION MANHOURS

Item of Work	Manhours Per Listed KW Rating		
	7500	10,000	12,650
Unload and hauling	300.	300.	350.
Plates	850.	850.	850.
Lift, set and align	7,650.	7,950.	7,950.
Piping	7,550.	7,900.	7,900.
Oil flush and test	2,250.	2,350.	2,350.
Electrical	1,900.	1,900.	1,900.
Insulation and lagging	2,200.	2,275.	2,275.
Start-up	650.	700.	700.
Total	23,350.	24,225.	24,275.

Item of Work	Manhours Per Listed KW Rating	
	15,625	16,500
Unload and hauling	375.	400.
Plates	950.	1,000.
Lift, set and align	8,700.	9,000.
Piping	8,600.	8,800.
Oil flush and test	2,450.	2,500.
Electrical	2,000.	2,000.
Insulation and lagging	2,425.	2,500.
Start-up	800.	800.
Total	26,300.	27,000.

See General Notes for explanation.

INERT GAS GENERATORS

MANHOURS REQUIRED EACH

Rated Output (SCFH Inerts)	Blower Horsepower	Approximate Weight Pounds	Manhours
2,000	1	1,200	25.0
3,000	2	1,500	30.0
5,000	3	2,000	35.0
8,000	3	2,500	38.0
12,000	7-1/2	3,000	42.0
25,000	10	4,000	50.0
40,000	20	6,500	60.0
60,000	25	7,500	65.0

Manhours include unloading, handling, hob hauling up to 2000 feet, rigging, picking, setting, aligning, and checking out of factory prefabricated unit consisting of generator tank body with burner and control panel, inert gas outlet plenum chamber, combustion blower and motor and interconnecting piping.

Manhours excludes installation of incoming and outgoing piping, electrical circuits and their connections.

GENERATORS—STANDBY

Skid-Mounted, Diesel Engine Driven

MANHOURS EACH

KW Rating	KVA Rating	Approximate Weight Pounds	Manhours
35	43.8	1,860	50.0
45	56.3	2,050	55.0
60	75.0	2,483	60.0
90	112.5	2,650	70.0
125	156.3	5,025	90.0
200	250.0	8,110	130.0
250	312.5	7,175	120.0

The following are common to all the above electrical systems:

Engine RPM	—	1800
Hertz	—	60
Phase	—	3
Power Factor	—	0.80

Manhours include unloading, handling, job hauling up to 2000 feet, rigging, picking, setting, aligning, hook-up, and checking out of skid-mounted, diesel driven electrical systems as outlined.

Manhours exclude installation of foundations or support structures.

HEAT EXCHANGERS
SHELL AND TUBE, FLOATING HEAD

INSTALLATION MANHOURS

Heating Surface Square Feet	MANHOURS		
	(1)	(2)	(3)
50	4.2	4.2	8.0
60	4.2	4.4	8.6
70	4.4	4.6	9.2
80	4.6	5.0	9.6
90	5.0	5.2	10.0
100	5.2	5.6	10.6
200	6.6	7.6	14.0
300	7.8	8.8	17.6
400	9.0	10.2	20.8
500	9.6	11.2	24.0
600	10.4	12.2	25.6
700	11.0	12.8	26.4
800	11.6	14.0	27.2
900	12.4	14.6	28.0
1000	12.8	15.2	28.8
2000	17.6	20.8	38.4
3000	19.2	25.6	44.8
4000	20.0	28.8	49.6
5000	24.0	30.4	57.7
6000	25.6	34.4	60.8

(1) Steel shell, steel tubes.

(2) Steel shell, copper tubes.

(3) Steel shell, stainless-steel tubes.

Manhours include unloading, handling, job hauling up to 2000 feet, rigging, picking, setting, aligning, and checking out of exchanger as outlined.

Manhours exclude installation of piping and connections.

HEAT EXCHANGERS
SHELL AND TUBE, FIXED-TUBE SHEET, U-TUBE

INSTALLATION MANHOURS

Heating Surface Square Feet	MANHOURS		
	(1)	(2)	(3)
50	3.2	4.0	5.6
60	3.6	4.2	6.0
70	3.8	4.8	6.4
80	4.0	5.0	6.8
90	4.2	5.2	7.4
100	4.4	5.4	7.6
200	5.4	7.2	10.4
300	6.4	8.4	12.6
400	7.2	9.6	14.4
500	7.8	10.4	16.0
600	8.2	11.2	17.6
700	9.0	12.0	19.2
800	9.2	12.8	20.8
900	9.6	13.6	22.4
1000	10.0	14.4	24.0
2000	13.2	20.0	30.4
3000	15.4	25.6	38.4
4000	17.6	27.2	41.6
5000	20.8	28.8	46.4
6000	22.4	30.4	51.2

(1) Steel shell, steel tubes.

(2) Steel shell, copper tubes.

(3) Steel shell, stainless-steel tubes.

Manhours include unloading, handling, job hauling up to 2000 feet, rigging, picking, setting, aligning, and checking out of exchanger as listed.

Manhours exclude installation of piping and connections.

HEAT EXCHANGERS—HAIRPIN TYPE

MANHOURS PER SECTION

Approximate Weight For Following Length For Various Section Types			Total Manhours For Following Lengths		
10'	20'	25'	10'	20'	25'
215#	350#	420#	19	23	29
400#	624#	750#	32	36	53
410#	670#	800#	33	40	56
515#	900#	1,060#	41	54	74
516#	1,000#	1,150#	45	60	81
600#	1,050#	1,225#	48	63	86
650#	1,100#	1,250#	52	66	88
—	—	1,300#	—	—	91

Manhours include unloading, handling, job hauling up to 2000 feet, rigging, picking, setting, aligning, and checking exchanger.

Manhours exclude installation of piping and connections.

HEAT EXCHANGERS–MISCELLANEOUS

INSTALLATION MANHOURS

Heating Surface Square Feet	MANHOURS			
	(1)	(2)	(3)	(4)
50	–	–	8.0	9.0
60	4.2	–	8.4	10.0
70	4.4	–	8.8	10.8
80	4.4	–	9.2	11.6
90	4.6	–	9.4	12.4
100	4.8	–	9.6	13.2
200	6.8	9.6	11.8	20.8
300	7.8	11.2	13.0	25.6
400	8.8	12.4	14.2	28.8
500	9.6	12.8	17.2	32.0
600	10.4	14.0	16.0	36.8
700	11.2	16.8	16.8	41.6
800	12.0	17.6	17.6	46.4
900	12.6	16.2	18.4	46.4
1000	13.2	16.6	19.2	48.0
2000	18.4	24.0	–	–
3000	13.2	27.2	–	–
4000	15.6	28.8	–	–
5000	–	30.4	–	–
6000	–	32.0	–	–

(1) Steel fin tubes.

(2) Steel reboilers.

(3) Jacketed pipe, steel.

(4) Jacketed pipe, glass and steel.

Manhours include unloading, handling, job hauling up to 2000 feet, rigging, picking, setting, aligning, and checking out of exchanger as outlined.

Manhours exclude installation of piping and connections.

EVAPORATORS-LONG TUBE VERTICAL

INSTALLATION MANHOURS

Heating Surface Square Feet	MANHOURS			
	Steel	Cast Iron Body Copper Tubes	Copper	Rubber-Lined Steel Karbate Tubes
100	–	–	–	1476.0
200	–	–	928.0	1760.0
300	–	488.0	992.0	2080.0
400	544.0	576.0	1072.0	2560.0
500	608.0	656.0	1104.0	2720.0
600	640.0	720.0	1152.0	2800.0
700	688.0	784.0	1184.0	2880.0
800	736.0	848.0	1232.0	3040.0
900	768.0	928.0	1264.0	3120.0
1,000	784.0	944.0	1360.0	3200.0
2,000	1024.0	1360.0	1472.0	4160.0
3,000	1216.0	1760.0	1584.0	–
4,000	1360.0	2160.0	–	–
5,000	1504.0	2400.0	–	–
6,000	1600.0	2560.0	–	–
7,000	1760.0	2720.0	–	–
8,000	1920.0	2880.0	–	–
9,000	2080.0	3040.0	–	–
10,000	2240.0	3248.0	–	–
20,000	2880.0	4640.0	–	–
30,000	3120.0	–	–	–
40,000	3440.0	–	–	–
50,000	3680.0	–	–	–

Manhours include unloading, handling, job hauling up to 2000 feet, rigging, picking, setting, aligning, and checking out of items as outlined.

Manhours exclude installation of supports and piping.

EVAPORATORS–HORIZONTAL TUBE

INSTALLATION MANHOURS

Heating Surface Square Feet	MANHOURS		
	(1)	(2)	(3)
100	384.0	—	—
200	528.0	1344.0	2,240.0
300	624.0	1440.0	2,560.0
400	720.0	1568.0	2,880.0
500	792.0	1680.0	3,040.0
600	864.0	1760.0	3,280.0
700	944.0	1920.0	3,520.0
800	976.0	2080.0	4,000.0
900	1040.0	2240.0	4,240.0
1000	1104.0	2400.0	4,480.0
2000	1544.0	3360.0	6,560.0
3000	1920.0	4720.0	8,960.0
4000	2240.0	5920.0	11,040.0
5000	2560.0	6720.0	12,800.0
6000	2800.0	7920.0	14,560.0
7000	2880.0	9120.0	16,800.0
8000	2960.0	9760.0	—

(1) Horizontal tube — cast iron body, copper tubes, steel.

(2) Forced circulation — cast iron body, copper tubes.

(3) Forced circulation — nickle, cast iron body, nickle tubes.

Manhours include unloading, handling, job hauling up to 2000 feet, rigging, picking, setting, aligning, and checking out of item as outlined.

Manhours exclude installation of supports and piping.

EVAPORATORS
JACKETED, GLASS-LINED STEEL VESSELS

INSTALLATION MANHOURS

Capacity Gallons	Manhours
50	304.0
60	336.0
70	352.0
80	368.0
90	384.0
100	400.0
200	496.0
300	576.0
400	616.0
500	648.0
600	704.0
700	752.0
800	784.0
900	816.0
1000	848.0

Manhours include unloading, handling, job hauling up to 2000 feet, rigging, picking, setting, aligning, and checking out of items as outlined.

Manhours exclude installation of supports and piping.

CONDENSERS-BAROMETRIC

INSTALLATION MANHOURS

Water Rate Gallons Per Minute	MANHOURS		
	Steel	Cast Iron	Rubber-Lined Steel
40	–	20.0	–
50	–	24.0	–
60	–	26.4	–
70	–	28.0	–
80	–	29.6	–
90	–	32.8	–
100	–	34.4	–
200	32.0	59.6	73.6
300	41.6	64.0	92.8
400	49.3	78.4	110.4
500	57.2	86.4	126.4
600	64.0	96.0	137.6
700	72.0	104.0	147.2
800	78.4	113.6	176.0
900	83.2	124.8	192.0
1000	92.8	148.0	208.0
2000	140.8	192.0	288.0
3000	192.0	256.0	352.0
4000	240.0	288.0	–
5000	272.0	320.0	–

Manhours include unloading, handling, job hauling up to 2000 feet, rigging, picking, setting, aligning, and checking out of items as outlined.

Manhours exclude installation of supports or other items.

VERTICAL TUBE SURFACE
CONDENSERS AND RECEIVERS

MANHOURS EACH

Cooling Water GPM @ 85° F	Manhours Each
20	14
30	15
55	16
65	22
110	23

Manhours include unloading, handling, job hauling up to 2000 feet, rigging, picking, setting, aligning, and checking items as listed.

Manhours exclude installation of supports or other items.

ACID COOLERS

MANHOURS EACH

Heat Exchanger Area Square Feet	Manhours Each
6.75	38
13.50	58
18.00	61
22.50	64
36.00	67
45.00	70
54.00	74
76.00	96
93.00	101
125.00	110
190.00	116
228.00	122

Manhours include unloading at job site, handling, hauling to project erection site within 2000 feet, rigging, picking, setting, aligning, and checking cooler including pump and motor.

Manhours exclude installation of support, piping connection, electrical circuit and connection.

FUEL OIL HEATERS

MANHOURS PER TON

Approximate Weight Tons From/To	Manhours Per Ton
0.0 / 0.5	12
0.5 / 0.7	15
0.7 / 1.0	24
1.0 / 1.5	35
1.5 / 1.9	44
1.9 / 2.7	63
2.7 / 3.0	70
3.0 / 3.5	82

Manhours include unloading, handling, job hauling up to 2000 feet, rigging, picking, setting, aligning, and checking items as outlined.

Manhours exclude installation of piping and piping tie-ins.

HEATING & VENTILATING UNITS

Truss or Suspended

MANHOURS PER UNIT

1000 btu Per Hour Capacity	Approximate Air Cu. Ft. Per Minute	Unit Weight In Pounds	Manhours
50	700	300	8.0
120	1,500	450	8.0
160	2,000	500	8.0
180	3,000	600	9.0
300	4,000	900	10.8
400	4,500	1500	18.0
500	5,500	1600	19.2
750	8,500	2500	30.0
1000	11,000	2600	31.2
1250	14,000	3500	42.0
1500	17,000	3700	44.4
1750	19,000	4300	51.6
2000	22,000	4500	52.0

Manhours include unloading, handling, job hauling up to 2000 feet, rigging, picking, setting, aligning, and checking out of units to 20 feet high as outlined. Increase manhours ¼% for each foot above 20 feet.

Manhours exclude installation of hangers, supports, piping, electrical circuits, and their connections.

AIR-CONDITIONING UNITS

Self-Contained, Truss or Suspended Ceiling Type

MANHOURS PER UNIT

Refrigeration Tons	Air Cu. Ft. Per Minute	Air Conditioning Units In Pounds	Manhours
6	2000	2000	39.0
10	4000	3000	56.7
15	6000	3800	68.4
20	8000	4000	72.0
30	12000	4500	81.0
40	16000	6000	108.0
50	20000	8000	132.0

Manhours include unloading, handling, job hauling up to 2000 feet, rigging, picking, setting, aligning, and checking out of units up to height of 20 feet. Increase above manhours by ¼% for each foot above 20 feet.

Manhours exclude installation of hangers, supports, electric circuits and their connections.

AIR HANDLING UNITS—
SINGLE ZONE & MULTI-ZONE

MANHOURS REQUIRED EACH

Capacity CFM	Fan Horsepower	Conditioned Area Square Feet	Manhours
Single Zone Units			
1,000	1.0	1,000	12.6
2,500	3.0	2,200	14.0
6,000	5.0	4,000	16.8
14,000	10.0	14,000	17.5
24,000	20.0	24,000	31.5
30,000	25.0	30,000	42.0
Multi-Zone Units			
4,000	3.0	4,000	12.6
6,000	5.0	6,000	15.4
10,000	7.5	10,000	16.8
15,000	15.0	15,000	25.2
22,000	20.0	22,000	28.0
30,000	25.0	30,000	42.0

The above units are based on a 550 FPM coil face velocity and 3 inches fan static pressure and include insulated casing, fan section, cooling coil section with a 6-row aluminum fin coil and drain pan, heating coil section with a 2-row aluminum fin coil, filter section with replaceable filters, fan motor, variable speed drive, and vibration isolators. Multi-zone units include zoning damper section.

Manhours include receiving at job site, off-loading from carrier, moving within 50 feet of final location site, uncrating, setting, and aligning.

Manhours do not include installation of ductwork, water or steam piping, motor starter or power wiring, and scaffolding. See respective tables for these time frames.

SPLIT SYSTEM AIR-COOLED PACKAGES

MANHOURS REQUIRED EACH

Capacity Tons	Average Weight Short Tons	Manhours
3	0.28	9.0
5	0.38	11.4
7-1/2	0.45	12.9
10	0.80	17.2
15	1.05	22.9
20	1.60	32.9
25	1.85	37.7
30	2.25	48.7
40	2.60	57.3
50	3.55	68.1

Manhours are based on receiving at job site, preassembled packages consisting of one high-side condensing unit and one low-side air handler companion piece, matched and ready for field piping.

Manhours include off-loading from carrier, moving within 50 feet of final location site, uncrating, setting, and aligning of the condensing unit on a preinstalled concrete slab along the exterior structure wall and suspending the air handler from the ceiling up to 15 feet high.

Manhours do not include installation of concrete slab, connecting piping, refrigerant charge, or scaffolding. See respective tables for these time frames.

PACKAGED HERMETIC CENTRIFUGAL WATER CHILLERS WITH WATER-COOLED CONDENSERS

MANHOURS REQUIRED EACH

Capacity Tons	Power KW	Average Weight Short Tons	Manhours
100	95	3.80	37.8
130	115	3.95	38.6
150	130	4.90	39.5
200	165	5.75	40.3
250	215	6.40	42.8
300	250	7.75	43.8
350	285	8.05	44.7
400	345	9.00	45.7
500	395	10.00	47.9
600	465	11.00	48.9
800	620	15.00	52.9
1000	720	17.50	58.0

Manhours are based on receiving package type units delivered to job site and include off-loading from carrier, moving within 50 feet of final location site, uncrating, setting, aligning, starting, and checking.

Manhours do not include installation of water supply and return piping, instruments within the water piping, or electrical power wiring. See respective tables for these time frames.

Water-cooled packages include hermetic compressor, 460-volt motor, cooler, condenser, internal piping and wiring, purge units, gauges, controls, insulation, lubrication system, oil, and refrigerant charge.

PACKAGED RECIPROCATING WATER CHII
WITH AIR-COOLED CONDENSERS

MANHOURS REQUIRED EACH

Capacity Tons	Power KW	Average Weight Short Tons	Manhou
20	23	1.35	27.7
30	35	1.60	28.4
40	53	2.10	29.5
50	62	2.95	32.8
60	90	3.05	34.2
80	97	4.70	37.8
100	116	5.10	38.6

PACKAGED RECIPROCATING HERMETIC
CHILLERS WITH WATER-COOLED CONDI

MANHOURS REQUIRED EACH

Capacity Tons	Power KW	Average Weight Short Tons	Manh
15	14	0.80	21
30	29	1.00	21
50	48	1.75	22
60	59	1.80	32
80	76	2.70	33
100	93	3.25	3
120	112	3.35	3
150	154	4.00	3

Manhours are based on receiving package type units delivered to job site and
from carrier, moving within 50 feet of final location site, uncrating, setting, ali
checking.

Manhours do not include installation of water supply and return piping, ins
water piping, or electrical power wiring. See respective tables for these time frai

Air-cooled packages include direct expansion cooler, air-cooled condenser, co
for compressors and fan motors, operating and safety controls, insulation, interr
charge, and vibration eliminators.

Water-cooled packages include hermetic compressor, motor, cooler, condense
wiring, motor starters, insulation, operating and safety controls, refrigerant
eliminators.

ROOM FAN COIL UNITS

MANHOURS REQUIRED EACH

Capacity CFM	Cooling BTUH	Heating BTUH	Water GPM	Electric Watts	Weight Pounds	Manhours
200	6,400	17,000	2	70	85	2.40
300	9,150	26,100	2	90	100	2.60
400	12,650	32,300	3	110	110	2.80
600	17,660	44,700	4	130	130	3.00

The above units include insulated cabinet with discharge air grill, 3-row aluminum fin coil, 115-volt, 1-phase motor, fan, 3-speed fan switch, thermostat, manual change-over switch, two-way solenoid valve, and throwaway filter.

Manhours include receiving at job site, off-loading from carrier, moving within 50 feet of final location, uncrating, setting, and aligning inside of building at grade.

If unit is to be supported from existing structural framing, up to 15 feet above floor, increase manhours 25%.

Manhours do not include installation of ductwork, water or steam piping, motor starter, or power wiring. See respective tables for these time frames.

COMBINATION HEATING & COOLING UNITS

MANHOURS REQUIRED EACH

Furnace Capacity BTU/Hour	Gas BTU/Hour	Forced Air Fan Motor Horsepower	Cooling Capacity Tons	Nominal CFM	Unit Weight Pounds	Installation Manhours
80,000	100,000	1/2	3	1,550	400	10.0
100,000	125,000	1/2	4	1,800	460	12.0
120,000	150,000	3/4	5	2,400	500	16.0

The above units include an upflow type gas-fired furnace and an electrically operated cooling system consisting of an evaporator coil factory mounted in bonnet of furnace, and a remote located, air-cooled condensing unit containing compressor, condenser coil, propeller fan and motor, and electrical controls. Cooling components are factory precharged with refrigerant and are equipped to receive furnished precharged tubing. Furnaces are certified for installation on combustible flooring.

Manhours include receiving at job site, off-loading from carrier, moving within 100 feet of final location, setting and aligning furnace, mounting air-cooled condenser a maximum of 100 feet from furnace, and installing a wall-mounted line voltage thermostat.

Manhours do not include installation of gas piping, power wiring or interconnecting control wiring. See respective tables for these time requirements.

ROOF-MOUNTED HEATING & COOLING UNITS

COMBINATION UNITS

MANHOURS REQUIRED EACH

Cooling Capacity Tons	Heating Capacity BTU/Hour	Power KW	Gas BTU/Hour	Nominal CFM	Average Weight Short Tons	Manhours
5	112,500	8	150,000	2,000	0.50	18.0
7-1/2	165,000	12	225,000	3,000	0.60	19.2
10	206,000	16	275,000	4,000	1.10	20.1
15	270,000	19	360,000	6,000	1.15	21.0
20	360,000	25	480,000	8,000	1.80	24.0
30	540,000	40	720,000	12,000	2.65	25.6
40	675,000	50	900,000	16,000	3.20	26.8
50	835,000	62	1,115,000	20,000	3.65	28.0

COMBINATION MULTI-ZONE MODULAR UNITS

MANHOURS REQUIRED EACH

Cooling Capacity Tons	Quantity of Zone Modules	Heating Capacity BTU/Hour	Power KW	Gas BTU/Hour	Nominal CFM	Average Weight Short Tons	Manhours
20	8	360,000	21	480,000	8,000	2.18	39.2
25	10	450,000	28	600,000	10,000	2.30	44.8
30	12	540,000	34	720,000	12,000	3.00	50.4

Manhours are based on receiving preassembled package type units, prewired, prepiped, and charged with refrigerant.

Units are electrically cooled, gas heated, and wired for 230 or 460 volts, 3-phase 60 Hz.

Installation of separately shipped roof curbs for units of 20 tons or greater is included.

Manhours do not include installation of power wiring or gas piping. See respective tables for these time frames.

CHILLED WATER PUMPS

MANHOURS REQUIRED EACH

Pump GPM	Head Feet	Discharge Pipe Size	Motor Horsepower	Weight Pounds	Installation Manhours
300	46	2-1/2"	5	340	12.0
480	87	3"	15	545	16.9
1,200	58	6"	25	1,450	20.2
1,800	46	6"	30	1,500	21.1
2,400	46	8"	40	2,000	23.9
3,000	46	8"	50	2,025	24.6
3,750	55	8"	75	2,200	27.6
4,200	66	8"	100	2,300	30.0
4,800	58	10"	100	2,450	30.6
5,500	50	12"	100	2,800	30.6

The pumps are cast iron, single-stage, horizontal cradle-mounted, vertical split case, end suction, top discharge, 125-pound flat face flanged with dripproof type motors, 1,700 RPM, direct connected by use of a flexible coupling. Pump and motor are both mounted on a common base plate.

Manhours include job handling, hauling, rigging, and setting and aligning of pump and motor.

Manhours do not include installation of piping, instrumentation and electrical motor starter and hook-up. See respective tables for these time frames.

CONDENSER WATER PUMPS

MANHOURS REQUIRED EACH

Pump GPM	Head Feet	Discharge Pipe Size	Motor Horsepower	Weight Pounds	Installation Manhours
350	59	3″	7.5	480	13.8
600	69	4″	15	680	16.9
1,500	46	6″	25	1,450	20.2
2,200	50	8″	40	2,000	23.9
3,000	46	8″	50	2,025	24.6
3,700	56	8″	75	2,200	27.6
4,500	62	10″	100	2,450	30.6
5,250	53	12″	100	2,800	30.6
5,750	48	14″	100	4,000	35.0
6,000	46	14″	100	4,000	35.0

Pumps are cast iron, single-stage, horizontal cradle-mounted, vertical split case, end suction, top discharge, 125-pound flat face flanged with dripproof type motors, 1,750 RPM through 12-inch and 1,160 RPM for 14-inch. Pump and motor and both mounted on a common base plate.

Manhours include job handling, hauling, rigging, and setting and aligning of pump and motor.

Manhours do not include installation of piping, instruments and electrical motor starter, and hook-up. See respective tables for these time frames.

CIRCULATING-BOOSTER PUMPS

MANHOURS REQUIRED EACH

Pump GPM	Head Feet	Pipe Size Connections	Motor Horsepower	Weight Pounds	Installation Manhours
Standard					
15	6	3/4″	1/8	28	10.0
20	5.5	1″	1/8	28	10.0
25	4.5	1-1/4″	1/8	28	10.0
30	2	1-1/2″	1/8	28	10.0
High Velocity					
30	7	1″	1/8	43	10.0
35	4	1-1/4″	1/8	43	10.0
35	5	1-1/2″	1/8	43	10.0
High Head					
10	9	1-1/2″	1/8	51	10.0
10	10	2″	1/6	42	10.0
10	16	1″	1/3	43	10.0

The pumps are of all bronze construction and are in-the-line, centrifugal type. Dripproof type motors are integral direct connected to pumps, 1,750 RPM, 115-volt, 60 Hz., single-phase. Pipe connections have threaded companion flanges bolted on.

Manhours include job handling, hauling, rigging, setting and aligning of pump and motor.

Manhours do not include installation of piping, instrumentation, and electrical motor starter, and hook-up. See respective tables for these time frames.

PREFABRICATED HOT WATER STORAGE TANKS

MANHOURS REQUIRED EACH

Tank Size Inches	Capacity Gallons	Weight Pounds	Installation Manhours
18 x 60	65	125	1.20
18 x 72	80	150	1.26
24 x 63	115	200	1.33
24 x 75	140	240	1.40
30 x 66	190	360	1.58
30 x 78	225	420	1.73
30 x 90	260	480	1.88
36 x 82	325	535	1.95
36 x 94	380	605	2.16
36 x 106	430	680	2.24
36 x 118	485	750	2.40
42 x 96	535	990	2.98
42 x 108	600	1100	3.60
48 x 96	700	1110	3.60
48 x 108	800	1230	3.68

Hot water storage tanks are ASME code constructed, galvanized carbon steel at 125 PSI working pressure. Inlet and outlet connections up to 6 inches are included. Manholes are included on 42-inch and larger diameter tanks.

The weight of factory installed tank heating coils is not included and therefore must be added to the above weight, dependent on size, for total lifting weight.

Manhours include receiving at job site, off-loading from carrier, moving within 50 feet of final location, rigging, picking, setting, and aligning.

Manhours do not include steam, condensate, or water piping, insulation, or instrumentation. See respective tables for these time frames.

COMPRESSION TANKS

MANHOURS REQUIRED EACH

Tank Size Inches	Capacity Gallons	Weight Pounds	Installation Manhours
13 x 34-1/2	15	50	1.00
13 x 51	24	73	1.00
13 x 61	30	79	1.00
16-1/4 x 53	40	90	1.10
16-1/4 x 76-1/2	60	145	1.40
20-1/4 x 68	80	170	1.58
20-1/4 x 82	100	200	1.73
24-1/4 x 71-1/2	120	334	2.16
24-1/4 x 83-1/2	144	383	2.20
30 x 60	163	497	2.40

Compression tanks are ASME code construction, black carbon steel at 125 PSI working pressure and are prefabricated with gauge glass connections.

Manhours include receiving at job site, off-loading from carrier, moving within 50 feet of final location, rigging, picking, setting, and aligning.

Manhours do not include steam, condensate or water piping, insulation or instrumentation. See respective tables for these time requirements.

COOLING TOWERS FOR RECIPROCATING & CENTRIFUGAL CHILLERS

MANHOURS REQUIRED EACH

Refrigeration Capacity Tons	Motor Horsepower	Average Weight Short Tons	Manhours
20	3/4	1.00	40.0
30	1-1/2	1.11	43.0
40	1-1/2	1.20	55.0
50	2	1.30	61.0
65	3	1.50	76.0
85	3	2.00	96.0
100	5	2.25	122.0
200	10	2.90	244.0
300	15	4.00	330.0
400	20	5.00	413.0
600	25	6.50	558.0
800	30	8.00	698.0
1000	50	12.50	824.0

Manhours are based on receiving prefabricated units, knocked down, and delivered to job site and include off-loading from carrier, moving within 50 feet of final location site, field assembling, erecting, aligning, and anchoring at ground elevation.

If tower is to be set on structure roof, increase manhours 5%.

Manhours do not include installation of concrete basin, supporting steel, grillage, water treatment system, condenser water piping, motor starters, or power wiring. See respective tables for these time frames.

FANS

INSTALLATION MANHOURS

Capacity Cubic Feet Per Minute	MANHOURS		
	Propeller	Venexal	Centrifugal
1,000	2.4	—	—
2,000	2.8	—	—
3,000	3.0	7.4	4.8
4,000	3.6	7.6	6.0
5,000	4.2	7.6	7.6
6,000	4.6	7.8	8.6
7,000	5.2	8.0	9.6
8,000	5.6	8.2	10.8
9,000	6.0	8.4	12.0
10,000	6.6	8.6	13.0
15,000	8.8	—	—
20,000	—	10.2	24.0
30,000	—	12.8	32.0
40,000	—	16.0	41.6
50,000	—	—	48.8
60,000	—	—	57.6
70,000	—	—	64.0
75,000	—	—	68.8

Manhours include unloading, handling, job hauling up to 2000 feet, rigging, picking, setting, aligning, and checking out of fans as outlined.

Manhours exclude installation of supports, electrical circuits, and their connections.

FANS AND MOTORS

Truss or Suspended

MANHOURS PER UNIT

Air Cu. Ft. Per Minute	Approximate Horsepower	Fan and Motor Weight In Pounds	Manhours
1,000 – 2,000	1 or less	400	6.0
3,000	1 1/2	500	7.5
4,000	2	600	9.0
6,000	3	700	10.5
10,000	5	1300	15.6
15,000	7 1/2	1800	21.6
22,000	10	2500	30.0
33,000	15	3900	42.0
40,000	20	5000	52.5
50,000	25	6000	63.0

Manhours include unloading, handling, job hauling up to 2000 feet, rigging, picking, setting, aligning and checking out of fans and motors to 20 feet high as outlined. Increase manhours by ¼% for each foot of height above 20 feet.

Manhours exclude installation of supports, hangers, electrical circuits, and their connections.

HOIST—OVERHEAD ELECTRIC

MANHOURS EACH

Lifting Capacity Tons	Manhours	Lifting Capacity Tons	Manhours
1/4	18	6	34
1/2	18	7-1/2	34
1	18	10	34
1-1/2	18	12	56
2	18	15	56
2-1/2	20	20	64
3	26	25	64
4	26	30	64
5	26	–	–

Manhours include unloading, handling, job hauling up to 2000 feet, assembling, rigging, picking, setting, aligning, and checking out of light, medium, and heavy duty hoist as outlined.

Manhours exclude installation of runway beams, electric circuits, and electrical connection to power source.

OVERHEAD TRAVELING BRIDGE CRANES

MANHOURS EACH

Crane Capacity Tons	Span Feet	Weight Pounds		Manhours Each	
		Floor Operated	Cab Operated	Floor Operated	Cab Operated
3	20	8,750	10,200	96	144
	30	10,600	12,100	101	151
	40	15,350	16,800	144	192
	50	20,600	22,000	151	202
	60	31,700	33,200	192	240
	75	35,000	36,500	202	252
5	20	9,000	10,500	96	144
	30	11,750	13,250	101	151
	40	18,350	19,800	144	192
	50	23,700	25,150	151	202
	60	32,000	33,500	192	240
	75	35,600	36,750	202	252
7½	20	9,700	11,000	96	144
	30	13,800	15,250	101	151
	40	18,450	20,200	144	192
	50	28,400	30,200	151	202
	60	33,000	34,800	192	240
	75	38,000	37,900	202	252
10	20	12,000	13,500	134	202
	30	15,600	17,100	141	212
	40	20,750	22,500	202	269
	50	30,600	32,400	212	282
	60	38,500	40,400	269	336
	75	49,000	51,000	274	353
15	20	20,200	22,000	134	202
	30	22,600	24,400	141	212
	40	24,800	26,600	202	269
	50	37,500	38,800	212	282
	60	48,000	49,400	269	336
	75	51,000	58,800	274	353
20	20	26,900	28,700	134	202
	30	29,350	31,130	141	212
	40	31,500	33,300	202	269
	50	44,650	46,000	212	282
	60	45,000	46,800	269	336
	75	58,400	60,200	274	353

Manhours include uploading at job site, handling, job hauling up to 2000 feet, assembling, rigging, picking, setting, aligning, and checking.

Manhours exclude installation of crane rails, electric circuits, and electrical connection to power source.

MIXERS—PROPELLER

INSTALLATION MANHOURS

Rating Horsepower	MANHOURS	
	Portable Mounted	Fixed Mounted
0.3	7.8	—
0.4	7.8	—
0.5	8.0	—
0.6	8.2	—
0.7	8.4	—
0.8	8.8	—
0.9	9.2	—
1.0	9.2	—
2.0	11.6	—
3.0	14.0	—
4.0	16.0	—
5.0	18.4	28.0
6.0	21.6	29.6
7.0	24.0	30.4
8.0	25.6	32.0
9.0	—	32.8
10.0	—	33.2
20.0	—	54.4
30.0	—	76.8
40.0	—	97.6
50.0	—	121.6

Manhours include unloading, handling, job hauling up to 2000 feet, rigging, picking, setting, aligning, and checking out of mixer as outlined.

Manhours exclude installation of electrical circuits and connections.

Portable-mounted mixer manhours are for direct or gear drive, vertical, clamp-mounted units.

Fixed-mounted mixer manhours are for direct or gear, vertical, plate, or flanged-mounted units.

For side entering, belt driven, horizontal units increase manhours 20%.

MIXERS-BLENDER TYPE

INSTALLATION MANHOURS

Capacity Cubic Feet	MANHOURS	
	Rotary Blender	Ribbon Blender
7	–	80.0
8	56.0	81.6
9	56.0	82.4
10	56.0	83.2
20	57.6	94.4
30	60.0	97.6
40	62.4	107.2
50	64.0	110.4
60	65.6	113.6
70	67.2	118.4
80	68.8	123.2
90	70.4	128.0
100	72.0	129.6
200	92.8	160.0
300	–	208.0

Manhours include unloading, handling, job hauling up to 2000 feet, rigging, picking, setting, aligning, and checking out of blender mounted on structural steel supports above open tank.

For mounting in enclosed tanks increase manhours 10%.

Manhours exclude installation of electric circuits and connections.

MIXERS-PAN AND SIGMA

INSTALLATION MANHOURS

Capacity Cubic Feet	MANHOURS	
	Pan Mixer	Sigma Mixer
2	–	208.0
3	–	212.8
4	96.0	219.2
5	107.2	224.0
6	113.6	230.4
7	124.8	236.8
8	130.4	243.2
9	140.8	249.6
10	145.6	256.0
20	208.0	320.0
30	256.0	–
40	304.0	–

Manhours include unloading, handling, hauling up to 2000 feet, rigging, picking, setting, aligning, and checking out of unit.

Manhours exclude installation of electrical circuits and connections.

MIXERS—HOMOGENIZERS

MANHOURS REQUIRED EACH

Rating Horsepower	Pipeline Mixer	Batch Mixer
3	19.0	21.0
5	–	27.9
7-1/2	23.7	30.7
10	29.3	37.5
15	–	41.3
20	32.7	43.4
25	–	44.7
30	36.0	–

Manhours include unloading, handling, job hauling up to 2000 feet, rigging, picking, setting, aligning, and checking out of unit.

Manhours exclude installation of electrical circuits and connections.

Pipeline homogenizer mixer is a basic unit that employs a single, high-speed turbine starter assembly.

Batch homogenizer mixer is a high-speed, high-shear batch mixer used for mixing, emulisifying, and dispersing.

MIXERS—HIGH-INTENSITY

And Mixer-Cooler Combinations

MANHOURS REQUIRED EACH

Item Description	Approximate Weight Pounds	Mixer Motor Horsepower	Cooler Motor Horsepower	Cooling Surface Square Feet	Manhours
High Intensity Mixers					
Mixer	2,500	30/15	–	–	20.0
Mixer	3,500	50/25	–	–	22.4
Mixer	4,500	60/30	–	–	24.0
Mixer	5,500	100/50	–	–	25.6
Mixer	12,000	200/100	–	–	33.6
Mixer	16,500	400	–	–	39.6
Mixer-Cooler Combination					
Mixer with Horizontal Cooler	8,000	50/25	10	31	67.2
Mixer with Horizontal Cooler	11,000	100/50	20	54	79.2
Mixer with Horizontal Cooler	20,000	200/100	30	90	250.0
Mixer with Horizontal Cooler	26,000	400	50	140	650.0

Manhours include unloading, handling, job hauling up to 2000 feet, rigging, picking, setting, aligning, and checking out of all components.

Manhours exclude the installation of electrical circuits and connections.

High-intensity mixers offer high shear and impact action when applied to any material that flows or can be fluidized.

PUMPS—CENTRIFUGAL

MANHOURS EACH

Prime Mover or Motor Horsepower Range	Single Stage		Two Stage		Multi- Stage		Vertical In Line Single Stage	
	1	2	1	2	1	2	1	2
0 - 15	3.00	20.0	3.30	22.0	3.60	24.0	3.00	20.0
16 - 30	2.50	50.0	2.75	55.0	3.00	60.0	2.50	50.0
31 - 50	2.00	75.0	2.20	8.25	2.40	90.0	2.00	75.0
51 - 75	1.75	110.0	1.90	121.0	2.10	132.0	1.75	110.0
76 - 100	1.50	150.0	1.65	165.0	1.80	180.0	1.50	150.0
101 - 125	1.50	155.0	1.65	170.0	1.80	186.0	1.50	155.0
126 - 300	1.25	200.0	1.38	220.0	1.50	240.0	–	–
301 - 500	–	–	0.90	270.0	0.98	295.0	–	–
501 - 5,000	–	–	–	–	0.60	310.0	–	–
5,001 - 7,500	–	–	–	–	0.10	510.0	–	–

Code:
1—Unit Manhours Per Horsepower.
2—Minimum Manhours Per Pump.

Manhours include unloading, handling, job hauling up to 2000 feet, rigging, picking, setting, aligning, and checking out of pump as outlined.

Manhours exclude installation of incoming or outgoing piping, electrical circuits, and their connections.

Single stage, two-stage and multi-stage pumps are either cast iron bronze fitted or all iron, horizontal cradle or foot-mounted, vertical or horizontal split cage, mounted on steel base plates with coupling and coupling guard and packed stuffing box. Pumps are for pumping for general use, processing materials, water, water flooding, boiler feed, utility, pipeline descaling, and similar applications with a capacity of 50 to 8,000 gallons per minute and weigh 200 to 7,400 pounds each.

Vertical in-line single stage pumps are of cast iron bronze fitted or all iron construction with packed stuffing box and TEFC motors. Pumps are for general service use with a capacity of 50 to 800 gallons per minute and weigh 160 to 2,360 pounds each.

PUMPS—VERTICAL TURBINE AND SUMP

MANHOURS EACH

Motor Horsepower Range	Vertical Turbine Pumps						Single Stage Sump Pumps	
	Single Stage		Two Stage		Multi Stage			
	1	2	1	2	1	2	1	2
0 - 15	3.6	24.0	4.00	26.0	4.30	29.0	4.50	30.0
16 - 30	3.0	60.0	3.30	66.0	3.60	72.0	3.75	75.0
31 - 50	2.4	90.0	2.60	99.0	2.90	108.0	3.00	112.0
51 - 75	2.1	132.0	2.30	145.0	2.50	158.0	2.60	165.0
76 - 100	1.8	180.0	2.00	198.0	2.20	216.0	–	–
101 - 125	1.8	186.0	2.00	205.0	2.20	223.0	–	–
126 - 300	1.5	240.0	1.65	264.0	1.80	316.0	–	–
301 - 500	0.9	280.0	1.00	308.0	1.10	336.0	–	–
501 - 600	0.7	360.0	0.75	396.0	0.80	432.0	–	–

Code:
1—Unit Manhours Per Horsepower.
2—Minimum Manhours Per Pump.

Manhours include unloading, handling, job hauling up to 2000 feet, rigging, picking, setting, aligning, and checking out of pump as outlined.

Manhours exclude installation of incoming or outgoing piping, electrical circuits, and their connections.

Vertical turbine sump pumps are of cast iron bronze fitted construction with semi-open or enclosed impellers and packed stuffing box and are wet type. Pumps are for water supply, cooling towers and process liquids use with a capacity of 50 to 10,000 gallons per minute and weigh 250 to 5,700 pounds each.

Vertical single stage sump pumps are of cast iron bronze fitted or all iron construction with semi-open impellers and cast iron suction strainers. Pumps are premounted on a steel support plate. These pumps have a capacity of 50 to 2,000 gallons per minute and weigh 350 to 2,030 pounds each.

PUMPS—POWER AND INTERNAL GEAR ROTARY

MANHOURS EACH

Motor Horsepower Range	Power Pumps				Internal Gear Rotary Pumps			
	Light Duty		Heavy Duty		General Purpose		Heavy Duty	
	1	2	1	2	1	2	1	2
0 - 15	1.50	10.0	1.80	12.0	2.25	15.0	2.50	18.0
16 - 30	1.25	25.0	1.50	30.0	1.88	38.0	2.00	40.0
31 - 50	1.00	37.5	1.20	45.0	1.25	56.0	–	–
51 - 75	0.90	55.0	1.10	66.0	–	–	–	–
76 - 100	0.75	75.0	0.90	90.0	–	–	–	–
101 - 125	–	–	0.90	94.0	–	–	–	–
126 - 300	–	–	0.78	120.0	–	–	–	–

Code:

1—Unit Manhours Per Horsepower

2—Minimum Manhours Per Pump.

Manhours include unloading, handling, job hauling up to 2000 feet, rigging, picking, setting, aligning, and checking out of pump as outlined.

Manhours exclude installation of incoming or outgoing pipe, electrical circuits, and their connections.

Light-duty power pumps include forged steel pump cylinder, 410 stainless steel plungers, valves and seats, and are mounted on a fabricated steel base plate with V-belt drive and belt guard attached. These pumps are designed for petroleum and industrial applications with a capacity of 10 to 170 gallons per minute and weigh 850 to 2,500 pounds each.

Heavy-duty power pumps are of the direct connection type and are designed for petroleum, oil field, chemical, petrochemical, hydraulic, and industrial applications. Pumps are mounted on fabricated steel base plates and consist of forged steel pumphead, integral gear and pinion with hardened 410 stainless steel plungers, valves and seats. These pumps are capable of pumping 6 to 500 gallons per minute and weigh 2,450 to 13,880 pounds each.

Internal gear rotary pumps are of cast iron construction, mounted on steel base plates and are for handling thin liquids. Pumps are either V-belt driven with semi-enclosed guard or direct motor connected. These pumps are capable of pumping 1.5 to 450 gallons per minute and weigh 60 to 830 pounds each.

PUMPS—VACUUM
High Vacuum, Two-Stage, and Multi-Stage

MANHOURS EACH

Motor Horsepower			Approximate Weight in Pounds	Manhours
First Stage	Second Stage	Third Stage		
High Vacuum				
1-1/2	—	—	315	15.0
2	—	—	345	16.0
3	—	—	565	18.0
7-1/2	—	—	950	24.0
10	—	—	1,750	26.0
2-10's	—	—	3,800	28.0
30	—	—	5,500	30.0
Two-Stage Vacuum				
2	7-1/2	—	1,350	26.0
2	10	—	2,300	28.0
Multi-Stage Vacuum				
3	—	7-1/2	1,685	26.0
7-1/2	—	7-1/2	2,700	28.0
7-1/2	—	10	3,500	30.0
10	—	7-1/2	2,800	30.0
10	—	10	3,600	34.0
20	—	10	5,250	40.0
25	—	10	5,400	42.0
30	—	2-10's	7,900	48.0
30	—	30	9,600	52.0

Manhours include unloading, handling, job hauling up to 2000 feet, rigging, picking, setting, aligning, and checking out of pump as outlined.

Manhours exclude installation of incoming and outgoing piping, electrical circuits, and their connections.

Vacuum pumps are complete units with motors, V-belt drives, and motor guards.

PUMPS—SEWAGE NONCLOG

MANHOURS EACH

Prime Mover or Motor Horsepower Range	Horizontal		Vertical	
	1	2	1	2
0 - 10	2.00	15.0	2.70	20.0
11 - 20	1.75	25.0	2.30	35.0
21 - 30	1.50	40.0	–	–
31 - 40	1.35	50.0	–	–
41 - 50	1.25	60.0	–	–
51 - 60	1.00	65.0	–	–

Code:

1—Unit Manhours Per Horsepower.

2—Minimum Manhours Per Pump.

Manhours include unloading, handling, job hauling up to 2000 feet, rigging, picking, setting, aligning, and checking out of pump as outlined.

Manhours exclude installation of incoming or outgoing piping, electrical circuits, and their connections.

Horizontal nonclog sewage pumps are of the end suction, top discharge type and are of cast iron construction, steel base plate mounted with coupling, coupling guard and packed stuffing box. Pumping capacity range is 50 to 2,800 gallons per minute and pumps weigh 250 to 990 pounds each.

Vertical nonclog sewage pumps are mounted on steel support plates and are of cast iron bronze fitted or all iron construction with 410 stainless steel shaft. Pumping capacity range is 50 to 2,800 gallons per minute and pumps weigh 385 to 910 pounds each.

CHEMICAL METERING PUMPS

MANHOURS EACH

GPH Per Head	Discharge Pressure PSI	Motor HP	Manhours Each		
			1	2	3
7.50	860	⅓	19	—	—
1.00	2,000	½	22	—	—
10.00	1,350	1	27	—	—
60.00	500	1½	30	—	—
20.80	150	¼	—	20	—
41.60	150	¼	—	20	—
20.80	100	¼	—	18	—
105.00	50	¼	—	18	—
0.86	700	⅓	—	—	19
2.70	1,100	⅓	—	—	19
7.00	925	⅓	—	—	19
18.00	350	⅓	—	—	19

1. Simplex chemical feed pump, 316 stainless steel liquid end construction, with double ball checks on suction and discharge, and spring-loaded, chevron-type packing.
2. Simplex or duplex chemical feed pumps, with Hypalon®-diaphragm and suction, discharge and injection assemblies included.
3. Simplex or duplex chemical metering pumps, with high-pressure, hydraulically balanced Teflon® diaphragm, oil-immersed operating parts, and micrometer-type feed rate adjustment.

Manhours include unloading, handling, job hauling up to 2000 feet, rigging, picking, setting, aligning, and checking pumps as outlined.

Manhours exclude installation of incoming and outgoing piping, electrical circuits, and their connections.

SCALES—TRUCK

Mechanical Lever Systems

MANHOURS EACH

Scale Capacity Tons	Platform Size Length X Width Feet	Approximate Total Weight Pounds	Manhours
10	18 x 10	5,500	200.0
15	22 x 10	6,500	240.0
15	30 x 10	8,000	300.0
20	10 x 10	5,000	200.0
25	24 x 10	7,000	240.0
25	34 x 10	10,000	320.0
30	10 x 10	5,000	200.0
30	12 x 10	6,200	230.0
30	24 x 10	8,500	320.0
30	34 x 10	11,000	352.0
50	45 x 10	13,000	368.0
50	50 x 10	14,000	400.0
50	60 x 10	16,500	410.0
50	70 x 10	20,000	496.0
60	50 x 10	15,000	410.0
60	60 x 10	18,000	475.0
60	70 x 10	22,000	500.0
80	60 x 10	18,500	480.0
80	70 x 10	23,500	532.0

Manhours include unloading, handling, job hauling up to 2000 feet, rigging, picking, assembling, setting, aligning, and checking out of all scale components for scales as outlined above.

Manhours exclude pit construction and installation of platform.

Scales are for weighing any highway type vehicle mechanically.

SCALES—TRUCK

Electronic Load Cell System

MANHOURS EACH

Scale Capacity Tons	Platform Size Length X Width Feet	Approximate Total Weight Pounds	Manhours
20	10 x 10	2,800	150.0
30	10 x 10	3,100	160.0
30	12 x 10	3,500	202.0
50	45 x 10	10,000	305.0
50	50 x 10	12,000	360.0
50	60 x 10	15,500	426.0
60	60 x 10	17,000	470.0
60	70 x 10	19,000	485.0
75	60 x 10	20,500	504.0
75	70 x 10	22,000	521.0

Manhours include unloading, handling, job hauling up to 2000 feet, rigging, picking, assembling, setting, aligning, and checking out of all scale components for scales as outlined above.

Manhours exclude pit construction and installation of platform and main electrical power source.

Scales measure weight electronically rather than mechanically.

SCALES—INDUSTRIAL
Built-In Type

MANHOURS EACH

Scale Capacity Tons	Platform Size Length X Width Feet	Approximate Total Weight Pounds	Manhours
5	6 x 5	3,300	80.0
5	8 x 6	3,900	96.0
5	9 x 7	4,500	104.0
10	8 x 6	4,100	98.0
10	9 x 7	4,700	110.0
15	9 x 7	5,300	120.0
15	10 x 10	6,600	130.0

Manhours include unloading, handling, job hauling up to 2000 feet, rigging, picking, assembling, setting, aligning, and checking out of all scale components for scales as outlined.

Manhours exclude pit construction.

Mechanical type scales are for heavy service use such as warehousing, industrial plants, bulk plants, and mines.

SCALES—AUTOMATIC BAGGING
Gross and Net Bagging

MANHOURS EACH

Type of Material to Be Handled	Feed Supply Method	Approximate Weight Pounds	Manhours
GROSS BAGGING			
Mechanically Operated			
Free Flowing	Gravity Fed	280	44.0
Semi-Free Flowing	Belt Fed	730	60.0
Powders	Screw Fed	980	75.0
Electrically Operated			
Semi-Free Flowing	Belt Fed	1,575	60.0
Powders	Screw Fed	1,875	80.0
Extremely Cohesive Powders	Pigment Fed	2,075	100.0
NET BAGGING			
Mechanically Operated			
Dry, Free Flowing Granular	–	980	80.0
Wet, Sluggish	–	1,875	95.0
Sticky Sluggish	–	2,075	110.0
Electrically Operated			
Free Flowing, Non-Flushing	–	1,380	85.0
Semi-Free Flowing	–	2,075	120.0
Mixed Granular	–	2,075	120.0
Free Flowing Granular	–	1,480	90.0
Sticky Sluggish	–	3,475	130.0

Manhours include unloading, handling, job hauling up to 2000 feet, assembling, rigging, picking setting, aligning, and checking out of all scale components for scales as outlined.

Manhours exclude installation of foundations or supports and electrical circuits from power source.

SCALES—BULK WEIGHING

MANHOURS EACH

Single Discharge Capacity Pounds	Approximate Weight Pounds	Manhours
Group One		
60	750	48.0
120	960	48.0
360	2,050	60.0
600	2,475	60.0
4,000	3,060	75.0
6,000	3,425	75.0
Group Two		
70	475	40.0
200	1,375	50.0
300	1,575	55.0
500	3,250	75.0
600	3,350	75.0
Group Three		
2,000	8,575	72.0
10,000	19,450	120.0
30,000	42,450	144.0

Group One—Open type for free flowing granular materials.
Group Two—Dust enclosed type for handling any bulk materials.
Group Three—Dust enclosed type with automatic hopper scales for granular materials.

Manhours include unloading, handling, job hauling up to 2000 feet, assembling, rigging, picking, setting, aligning, and checking out of all scale components for scales as outlined.

Manhours exclude installation of foundations or structural supports.

CENTRIFUGAL SEPARATORS
BATCH TOP SUSPENDED

INSTALLATION MANHOURS

Diameter Inches	MANHOURS		
	Steel	Rubber-Covered Steel	Stainless Steel
20	45.4	94.0	194.4
24	61.4	123.5	259.2
30	77.4	159.0	324.0
36	97.8	200.0	397.0
40	113.0	243.0	469.8
42	118.4	259.2	492.2

Manhours include unloading, handling, job hauling up to 2000 feet, rigging, picking, setting, aligning, and checking out of items as listed.

Manhours exclude installation of support structures and electrical power source.

CENTRIFUGAL SEPARATORS
BATCH BOTTOM DRIVEN

INSTALLATION MANHOURS

Diameter Inches	MANHOURS		
	Steel	Rubber-Covered Steel	Stainless Steel
20	36.4	55.1	105.3
24	41.8	64.8	121.5
30	47.0	74.6	163.7
36	55.1	81.0	210.6
40	59.3	92.4	243.0
42	63.2	100.0	267.3
48	68.1	107.0	291.6

Manhours include unloading, handling, job hauling up to 2000 feet, rigging, picking, setting, aligning, and checking out of items as listed.

Manhours exclude installation of support structures and electrical power source.

CENTRIFUGAL SEPARATORS
BATCH AUTOMATIC

INSTALLATION MANHOURS

Diameter Inches	MANHOURS		
	(1)	(2)	(3)
18	127.8	194.4	—
20	145.8	243.0	421.2
24	194.4	291.6	486.0
28	243.0	356.4	615.6
30	269.4	421.2	—
40	340.2	550.8	—
50	421.2	680.4	—
60	486.0	777.6	—
70	567.0	858.6	—
80	615.6	1119.6	—

(1) Baker-Perkins Steel

(2) Baker-Perkins Stainless Steel

(3) Sharples-Super D Hydrator Stainless Steel

Manhours include unloading, handling, job hauling up to 2000 feet, rigging, picking, setting, aligning, and checking out of items as outlined.

Manhours exclude installation of support structures and electrical power source.

CENTRIFUGAL SEPARATORS
HIGH SPEED

INSTALLATION MANHOURS

Diameter Inches	MANHOURS	
	Tubular	Disk
4	96.0	—
5	160.0	—
6	256.0	—
8	—	—
10	—	—
12	—	144.0
14	—	192.0
16	—	240.0
18	—	288.0

Manhours include unloading, handling, job hauling up to 2000 feet, rigging, picking, setting, aligning, and checking out of items as outlined.

Manhours exclude installation of support structures and electrical power source.

VANE-TYPE SEPARATORS

MANHOURS EACH

Inlet & Outlet Line Size Inches	Flange Rating	Manhours Each
2	150#	4.4
2	300#	5.8
3	150#	5.8
3	300#	7.1
4	150#	8.6
4	300#	10.1
6	150#	10.2
6	300#	11.6
8	150#	12.1
8	300#	13.7
10	150#	14.0
10	300#	17.3
12	150#	17.3
12	300#	20.2

Manhours include unloading, handling, job hauling up to 2000 feet, rigging, picking, setting, aligning, and checking items as outlined.

Manhours exclude installation of incoming or outgoing piping.

VIBRATING SEPARATORS

MANHOURS EACH

Diameter of Unit Inches	Number of Decks	Motor Horsepower	Approximate Weight Pounds	Manhours Each
18	4	¼	163	15
24	4	⅓	270	19
30	3	½	466	24
48	4	2½	971	29
60	3	2½	975	34
72	2	5	2,100	38

Note: Separators can be constructed of carbon steel or stainless steel material.

Electrical motors are totally enclosed, 230/460 or 575 v; 60 cycles; 3-phase; 1200 rpm.

Manhours include unloading, handling, job hauling up to 2000 feet, rigging, picking, setting, aligning, and checking separators.

Manhours do not include installation of electrical circuits and their connection.

API TYPE OIL/WATER SEPARATORS

MANHOURS EACH

Separator Tank Size Dia. × Length	Weight Pounds	Manhours Each
4.5 × 20.5	4,600	29
8.0 × 28.0	14,000	36
8.0 × 38.0	16,500	38

Manhours include unloading, handling, job hauling up to 2000 feet, rigging, picking, setting, aligning, and checking items as outlined.

Manhours exclude installation of piping, piping connections, and concrete pad if required.

SIZE REDUCTION-CRUSHERS

INSTALLATION MANHOURS

Drive Horsepower	MANHOURS				
	Rotary	Sawtooth	Jaw	Crushing Rolls	Gyratory
2	16.0	–	–	–	–
3	20.0	–	48.0	–	–
4	24.0	–	56.0	–	–
5	25.6	32.0	60.8	128.0	–
6	27.2	35.2	65.6	136.0	–
7	28.8	38.4	70.4	140.8	–
8	30.4	41.6	75.2	145.6	–
9	32.0	44.8	80.0	152.0	–
10	34.4	46.4	81.3	158.0	–
20	48.0	60.8	112.0	200.0	–
30	–	72.0	116.0	240.0	288.0
40	–	80.0	133.6	264.0	320.0
50	–	–	168.0	280.0	352.0
60	–	–	176.0	296.0	384.0
70	–	–	192.0	–	416.0
80	–	–	208.0	–	448.0
90	–	–	224.0	–	456.0
100	–	–	240.0	–	464.0
200	–	–	–	–	608.0
250	–	–	–	–	656.0

Manhours include unloading, handling, job hauling up to 2000 feet, assembling when necessary, rigging, picking, setting, aligning, and operational check-out.

Manhours exclude installation of electrical circuits and connections.

SIZE REDUCTION–MILLS, CUTTERS, PULVERIZERS

INSTALLATION MANHOURS

Drive Horsepower	MANHOURS				
	Attrition Mills	Swing Hammer Mills	Rotary Knife Cutters	Miko Pulverizer	Roller Mills
4	–	24.0	–	–	–
5	27.2	28.0	28.8	–	–
6	28.8	29.6	31.2	140.8	–
7	30.4	32.0	34.4	164.0	–
8	32.0	36.8	40.0	152.0	–
9	33.6	39.2	42.4	160.0	–
10	35.2	41.6	45.6	168.0	–
20	50.4	65.6	73.6	224.0	–
30	64.0	91.2	96.8	–	448.0
40	75.2	112.0	121.6	–	512.0
50	83.2	128.0	144.0	–	560.0
60	92.8	144.0	160.8	–	608.0
70	99.2	160.0	192.0	–	640.0
80	108.8	184.0	240.0	–	688.0
90	113.6	208.0	–	–	736.0
100	120.0	240.0	–	–	768.0
200	176.0	352.0	–	–	992.0
300	224.0	448.0	–	–	1184.0
400	–	576.0	–	–	1328.0

Manhours include unloading, handling, job hauling up to 2000 feet, assembling when necessary, rigging, picking, setting, aligning, and operational check-out.

Manhours exclude installation of electrical circuits and connections.

SIZE REDUCTION-BALL MILLS

INSTALLATION MANHOURS

Tons Per Hour Of Medium Hard Material	MANHOURS			
	(1)	(2)	(3)	(4)
1	–	–	496.0	800.0
2	–	432.0	656.0	1056.0
3	192.0	512.0	800.0	1232.0
4	240.0	608.0	928.0	1376.0
5	272.0	640.0	976.0	1520.0
6	288.0	704.0	1072.0	1600.0
7	320.0	768.0	1136.0	1760.0
8	352.0	800.0	1208.0	1920.0
9	368.0	848.0	1264.0	2080.0
10	384.0	896.0	1312.0	2240.0
20	576.0	1216.0	–	–
30	720.0	–	–	–
40	816.0	–	–	–
50	944.0	–	–	–

(1) 1-1/2-inch reduced to 10 mesh.

(2) 3/4-inch reduced to 48 mesh.

(3) 1/2-inch reduced to 100 mesh.

(4) 1/4-inch reduced to 98 per cent minus 325 mesh.

Manhours include unloading, handling, job hauling up to 2000 feet, assembling when necessary, rigging, picking, setting, aligning, and operational check-out.

Manhours exclude installation of electrical circuits and connections.

GRAVITY IMPACT CRUSHERS

MANHOURS EACH

Size Feet	Capacity Tons Per Hour	Manhours Each
9 × 12	500	518
9 × 16	650	544
9 × 20	830	571
9 × 24	970	600
12 × 16	800	691
12 × 20	940	726
12 × 24	1,070	762
12 × 28	1,200	800
14 × 20	1,120	960
14 × 24	1,270	1,008
14 × 28	1,400	1,058
14 × 32	1,500	1,111

Manhours include unloading, handling, job hauling up to 2000 feet, assembling when necessary, rigging, picking, setting, aligning, and operational check-out.

Manhours exclude installation of electrical motors, circuits, and connections.

SURGE ARRESTORS

MANHOURS EACH

Nominal Size Gallons	Fluid Port Connection	Design Operating Pressure PSI	Manhours Each
2½	3" 150# flg.	275	5
5	3" 150# flg.	275	5
10	3" 150# flg.	275	6
10	4" 150# flg.	275	6
25	4" 150# flg.	275	7
25	4" 300# flg.	500	7
40	4" 150# flg.	275	8
40	4" 300# flg.	500	8
80	4" 150# flg.	275	10
80	4" 300# flg.	500	10
100	4" 150# flg.	275	11
100	4" 300# flg.	500	11
120	4" 150# flg.	275	12
120	4" 300# flg.	500	12

Manhours include unloading, hauling, job hauling up to 2000 feet, rigging, picking, setting, aligning, and checking items as listed.

Manhours do not include installation of piping or bolt-ups.

THICKENERS–CONTINUOUS TYPE

INSTALLATION MANHOURS

Setting Area Square Feet	Manhours
100	128.0
200	134.4
300	142.4
400	150.4
500	158.4
600	168.0
700	176.0
800	192.0
900	208.0
1000	224.0
2000	256.0
3000	288.0
4000	304.0
4500	320.0

Manhours include unloading, handling, job hauling up to 2000 feet, rigging, picking, setting, aligning, and checking out of items as outlined.

Manhours exclude installation of piping, electrical circuits, and their connections.

VESSELS—PRESSURE

MANHOURS EACH

Weight Range Tons	Manhours Each
Horizontal Vessels	
0-5	90.0
6-10	120.0
11-20	200.0
21-30	250.0
31-40	325.0
41-60	420.0
61-100	550.0
101-150	625.0
151-200	750.0
201-250	830.0
251-300	900.0
Vertical Vessels (Towers)	
0-5	100.0
6-10	144.0
11-20	240.0
21-30	300.0
31-40	390.0
41-60	510.0
61-100	660.0
101-150	750.0
151-200	900.0
201-250	980.0
251-300	1,050.0

Manhours include unloading, handling, job hauling up to 2000 feet, erection study, rigging, picking, setting, and aligning of vessel as outlined.

Manhours exclude installation of trays, internals, packings, and inspection if required.

If tower is to be set with the use of poles, erection, and dismantling, time must be added.

Manhours are based on reasonable access to erection site. If vessel is to be erected in a congested area, this should be evaluated separately and an adjustment made to the manhours.

For inspection of trays or internals in refinery type columns add the following:
Remove and Replace Manway Cover . . . . 1.3 Manhours Per Tray
Check Tray and Tighten Retaining Bolts. . . . . 1.8 Manhours Per Tray

VESSELS—TRAY INSTALLATION

MANHOURS EACH

Column Diameter in Inches	MANHOURS EACH			
	1	2	3	4
36	6.0	6.6	7.4	9.3
42	7.5	9.0	10.2	11.4
48	9.6	10.4	12.6	14.1
54	10.4	13.5	15.0	17.1
60	13.5	15.9	18.3	20.1
66	15.0	18.3	20.7	22.8
72	17.4	21.0	23.4	26.1
78	19.8	23.4	26.7	29.7
84	22.5	27.0	30.3	33.6
90	25.2	30.3	34.2	38.1
96	27.6	33.3	37.5	41.4
102	30.6	36.3	40.8	45.6
108	33.6	40.2	45.6	50.1
114	36.3	44.1	50.4	54.6
120	39.3	47.1	53.1	59.1
126	42.6	50.7	57.6	63.9
132	45.6	54.6	61.5	68.4
138	48.6	58.5	65.4	72.9
144	51.6	62.1	69.9	77.4

Code:
  1—Single Downflow Valve or Perforated Type Trays.
  2—Double Downflow Valve or Perforated Type Trays.
  3—Single Downflow Bubble Cap Type Trays.
  4—Double Downflow Bubble Cap Type Trays.

Manhours include unloading, handling, job hauling up to 2000 feet, rigging, picking, setting, fastening, and aligning of trays passed through manway. An allowance for installation of seal pan under bottom tray is included.

Manhours exclude installation of vessel or other internals.

VESSELS—DEMISTING PADS

MANHOURS EACH

Vessel Diameter Inches	Weight—Pounds				Manhours Single Grid Installation
	1 Support	2 Pad	3 Grid	4 Total	
36	16	28	14	58	20.3
42	18	38	19	75	23.2
48	21	50	25	96	26.7
54	24	64	32	120	30.3
60	27	78	39	144	33.8
66	29	96	48	173	36.4
72	32	114	57	203	39.6
78	35	132	66	233	42.6
84	37	154	77	268	46.6
90	40	178	89	307	49.7
96	43	202	101	346	52.7
102	45	228	114	387	55.8
108	48	254	127	429	60.4
114	51	284	142	477	63.6
120	53	314	157	524	68.1
126	56	348	174	578	72.8
132	59	382	191	632	77.3
138	62	416	208	686	82.0
144	64	450	225	739	86.7

Weight Code:
1—Weight of One 1/4" x 2" Flat Bar Support Ring.
2—Weight of Pad.
3—Weight of Grid.
4—Total Weight for Bottom Grid.

Manhours include unloading, handling, job hauling up to 2000 feet, rigging, picking, setting, aligning, and fastening in place bottom grid as outlined.

If top and bottom grids are to be installed, add support and grid weight to above total weight and increase manhours 75%.

Manhours exclude installation of vessel or other internals.

VESSELS—TOWER PACKINGS

MANHOURS EACH

Vessel Diameter Inches	MANHOURS						
	1	2	3	4	5	6	7
12	–	4.0	–	4.0	–	8.0	6.0
18	–	5.0	–	5.0	–	10.0	8.0
24	–	6.0	8.0	6.0	–	12.0	8.0
30	–	7.0	9.0	7.0	–	14.0	10.0
36	–	8.0	10.0	8.0	–	16.0	12.0
48	8.0	–	12.0	12.0	8.0	–	12.0
60	12.0	–	16.0	16.0	12.0	–	16.0
72	16.0	–	20.0	20.0	16.0	–	20.0
84	18.0	–	24.0	24.0	18.0	–	24.0
96	22.0	–	28.0	28.0	22.0	–	28.0
108	24.0	–	32.0	32.0	24.0	–	32.0
120	28.0	–	36.0	36.0	28.0	–	36.0

Code:

1—Mult-Beam Support Plate.   5—Metal Weir Trough Distributor.
2—Metal Support Plate.   6—Metal "Weir Riser" or Orifice Distributor.
3—Metal Hold Down Plate.   7—Metal Redistributors.
4—Metal Bed Limiter.

For installation of pall rings, intalox saddles, or Raschig rings allow 0.75 manhours per cubic foot.

REMOVE AND REPLACE MANHOLE COVERS

MANHOURS EACH

Cover Size Inches	Manhours					
	150 lb. R.F.		300 lb. R.F.		600 lb. R.F.	
	1	2	1	2	1	2
14	10.0	16.0	12.0	19.0	13.0	20.0
16	13.0	20.0	16.0	24.0	16.0	25.0
18	16.0	24.0	19.0	29.0	20.0	30.0
20	19.0	28.0	23.0	34.0	24.0	35.0
24	21.0	32.0	25.0	38.0	26.0	40.0

Code:

1—Hinged Type.   2—Removable, Using Vessel Davit.

Manhours include unloading, handling, hauling up to 2000 feet, rigging, picking, setting, and aligning of listed items.

Manhours exclude installation of vessels or other components.

REACTORS–STEEL AGITATED, JACKETED

INSTALLATION MANHOURS

Capacity Gallons	MANHOURS		
	Steel 50 psi	Steel 300 psi	Steel 1500 psi
50	88.0	–	–
60	92.8	–	–
70	96.0	124.0	928.0
80	99.2	168.8	936.0
90	104.0	176.0	944.0
100	109.2	158.4	952.0
200	129.6	192.0	960.0
300	147.2	208.0	968.0
400	163.2	232.0	976.0
500	176.0	248.0	–
600	192.0	256.0	–
700	216.0	264.0	–
800	240.0	272.0	–
900	248.0	280.0	–
1000	256.0	288.0	–
2000	288.0	–	–

Manhours include unloading, handling, job hauling up to 2000 feet, rigging, picking, setting, aligning, and checking out of reactors as outlined.

Manhours exclude installation of incoming or outgoing piping, electrical circuits, and their connections.

REACTORS–AGITATED, JACKETED

INSTALLATION MANHOURS

Capacity Gallons	MANHOURS			
	(1)	(2)	(3)	(4)
50	120.0	152.0	–	–
60	128.0	160.0	–	–
70	136.0	168.0	256.0	1920.0
80	144.0	176.0	264.0	1984.0
90	152.0	184.0	272.0	2000.0
100	160.0	192.0	288.0	2016.0
200	208.0	256.0	480.0	2080.0
300	256.0	272.0	512.0	2112.0
400	272.0	296.0	560.0	2144.0
500	288.0	320.0	584.0	–
600	304.0	336.0	608.0	–
700	320.0	352.0	624.0	–
800	336.0	368.0	640.0	–
900	352.0	392.0	656.0	–
1000	368.0	416.0	672.0	–
2000	480.0	512.0	–	–

(1) Glass-lined steel 50 psi.

(2) Stainless steel 50 psi.

(3) Stainless steel 300 psi.

(4) Stainless steel 1500 psi.

Manhours include unloading, handling, job hauling up to 2000 feet, rigging, picking, setting, aligning, and checking out of reactors as outlined.

Manhours exclude installation of incoming or outgoing piping, electrical circuits, and their connections.

TANKS-VACUUM RECEIVER

INSTALLATION MANHOURS

Capacity Gallons	MANHOURS			
	Steel	Steel Jacketed	Stainless Steel	Stainless Steel Jacketed
30	16.0	20.8	32.0	57.6
40	17.6	22.4	35.2	60.0
50	19.2	24.0	37.6	63.2
60	20.8	24.8	40.8	65.6
70	22.4	25.6	42.4	67.2
80	24.0	26.4	44.0	68.8
90	25.6	27.2	45.6	70.4
100	27.2	28.8	46.4	72.8
200	28.0	31.2	57.6	83.2
300	28.8	33.6	64.0	92.8
400	29.6	36.8	70.4	98.4
500	30.4	29.2	75.2	104.0

Manhours include unloading, handling, job hauling up to 2000 feet, rigging, picking, setting, aligning, and checking out of tanks as outlined.

Manhours exclude installation of incoming or outgoing piping, electrical circuits, and their connections.

TANKS–VACUUM RECEIVER

INSTALLATION MANHOURS

Capacity Gallons	MANHOURS	
	Glass-Lined	Glass-Lined Jacketed
30	33.6	50.4
40	35.2	52.0
50	36.8	57.6
60	38.4	59.2
70	40.0	60.8
80	41.6	62.4
90	43.2	64.0
100	44.8	65.6
200	51.2	76.8
300	57.6	81.6
400	60.8	88.0
500	64.0	91.2

Manhours include unloading, handling, job hauling up to 2000 feet, rigging, picking, setting, aligning, and checking out of tanks as outlined.

Manhours exclude installation of incoming or outgoing piping, electrical circuits and their connections.

TANKS–AGITATED

INSTALLATION MANHOURS

Capacity Gallons	MANHOURS	
	Steel	Stainless Steel
100	89.6	176.0
200	112.0	224.0
300	128.0	256.0
400	144.0	288.0
500	152.0	304.0
600	161.6	320.0
700	176.0	336.0
800	192.0	352.0
900	208.0	368.0
1,000	224.0	384.0
2,000	272.0	480.0
3,000	288.0	560.0
4,000	304.0	608.0
5,000	312.0	648.0
6,000	352.0	720.0
7,000	384.0	752.0
8,000	416.0	784.0
9,000	432.0	800.0
10,000	448.0	832.0
20,000	536.0	1072.0
30,000	608.0	1216.0

Manhours include unloading shop fabricated segments welded together within shipping limits, handling, job hauling up to 2000 feet, rigging, picking, setting, aligning, field welding, and checking out of tanks as outlined.

Manhours exclude installation of agitator, incoming or outgoing piping, electrical circuits, and their connections.

TANKS–STORAGE

Redwood, Pine or Fir; Cypress; Lithcote-Lined Steel

INSTALLATION MANHOURS

Capacity Gallons	MANHOURS		
	Redwood Pine Or Fir	Cypress	Lithcote-Lined Steel
800	19.2	24.0	—
900	20.8	25.6	—
1,000	22.4	27.2	—
2,000	28.8	35.2	—
3,000	33.2	44.8	—
4,000	41.6	48.8	256.0
5,000	44.8	56.0	272.0
6,000	48.8	59.2	288.0
7,000	52.8	64.0	304.0
8,000	57.6	68.8	320.0
9,000	60.8	73.6	336.0
10,000	64.0	76.8	352.0
20,000	—	—	448.0
30,000	—	—	528.0
40,000	—	—	576.0
50,000	—	—	624.0
60,000	—	—	656.0
70,000	—	—	688.0
80,000	—	—	736.0
90,000	—	—	768.0
100,000	—	—	800.0

Manhours include unloading shop fabricated sections of sizes within shipping limits, handling, job hauling, rigging, picking, setting, fastening or welding, aligning, and checking out of tanks as outlined.

Manhours exclude installation of incoming or outgoing piping and their connections.

TANKS–STORAGE

Aluminum; Monel; Silver-Lined Steel
100 to 5000-Gallon Capacity

INSTALLATION MANHOURS

Capacity Gallons	MANHOURS		
	Aluminum	Monel	Silver-Lined Steel
100	60.8	105.6	144.0
200	83.2	144.0	240.0
300	99.2	168.0	288.0
400	113.6	192.0	320.0
500	128.0	216.0	384.0
600	144.0	240.0	432.0
700	156.8	248.0	464.0
800	163.2	256.0	480.0
900	176.0	264.0	512.0
1000	192.0	272.0	560.0
2000	272.0	400.0	808.0
3000	304.0	472.0	1040.0
4000	368.0	528.0	1248.0
5000	416.0	592.0	1624.0

Manhours include unloading shop fabricated sections of sizes within shipping limits, handling, job hauling up to 2000 feet, rigging, picking, setting, job welding, aligning, and checking out of tanks as outlined.

Manhours exclude installation of incoming or outgoing piping and their connections.

TANKS–STORAGE

Aluminum; Monel; Silver-Lined Steel
6000 to 100,000-Gallon Capacity

INSTALLATION MANHOURS

Capacity Gallons	MANHOURS		
	Aluminum	Monel	Silver-Lined Steel
6,000	448.0	624.0	1584.0
7,000	464.0	668.0	1760.0
8,000	496.0	736.0	1920.0
9,000	528.0	768.0	2080.0
10,000	560.0	784.0	2240.0
20,000	784.0	1056.0	—
30,000	944.0	1280.0	—
40,000	1104.0	1440.0	—
50,000	1216.0	1600.0	—
60,000	1296.0	1760.0	—
70,000	1440.0	1920.0	—
80,000	1520.0	2080.0	—
90,000	1600.0	2240.0	—
100,000	1760.0	2400.0	—

Manhours include unloading of shop fabricated sections of sizes within shipping limits, handling, job hauling up to 2000 feet, rigging, picking, setting, job welding, aligning, and checking out of tanks as outlined.

Manhours exclude installation of incoming or outgoing piping and their connections.

TANKS-STORAGE

Haveg; Copper; Glass-Lined Steel

INSTALLATION MANHOURS

Capacity Gallons	MANHOURS		
	Haveg	Copper	Glass-Lined Steel
100	–	49.6	99.2
200	–	65.6	121.6
300	–	78.4	131.2
400	–	89.6	144.0
500	–	97.6	155.6
600	–	108.8	160.0
700	–	113.6	168.0
800	–	121.6	176.0
900	–	128.0	192.0
1,000	152.0	134.4	208.0
2,000	208.0	184.0	–
3,000	256.0	240.0	–
4,000	288.0	256.0	–
5,000	320.0	280.0	–
6,000	352.0	296.0	–
7,000	400.0	312.0	–
8,000	–	344.0	–
9,000	–	360.0	–
10,000	–	376.0	–

Manhours include unloading shop fabricated sections of sizes within shipping limits, handling, job hauling up to 2000 feet, rigging, picking, setting, job welding, aligning, and checking out of tanks as outlined.

Manhours exclude installation of incoming or outgoing piping and their connections.

TANKS–STORAGE

Steel; Rubber-Lined Steel; Stainless-Clad Steel; Nickel Clad Steel Stainless Steel; Monel-Clad Steel; Iconell-Clad Steel

INSTALLATION MANHOURS

Capacity Gallons	MANHOURS			
	(1)	(2)	(3)	(4)
100	30.4	56.0	64.0	80.0
200	38.4	72.0	80.0	96.0
300	46.4	81.6	89.6	105.6
400	51.2	94.4	102.4	118.4
500	57.6	99.2	107.2	123.2
600	60.8	108.8	117.6	132.8
700	64.0	113.6	121.6	137.6
800	67.2	124.8	132.8	148.8
900	70.4	128.0	136.0	172.0
1,000	76.8	145.6	137.6	153.6
2,000	97.6	168.0	176.0	192.0
3,000	118.4	208.0	216.0	232.0
4,000	136.0	240.0	248.0	264.0
5,000	145.6	256.0	264.0	280.0
6,000	160.0	272.0	280.0	296.0
7,000	176.0	288.0	296.0	312.0
8,000	192.0	304.0	312.0	328.0
9,000	208.0	320.0	328.0	344.0
10,000	224.0	336.0	344.0	360.0

(1) Steel.

(2) Rubber-lined steel.

(3) Stainless-clad steel, nickle-clad steel.

(4) Stainless steel; monel-clad steel; inconel-clad steel.

Manhours include unloading shop fabricated sections of sizes within shipping limits, handling, job hauling up to 2000 feet, rigging, picking, setting, job welding, aligning, and checking out of tanks as outlined.

Manhours exclude installation of incoming or outgoing piping and their connections.

For steel cone roof and floating roof tanks increase manhours 20 and 25% respectively.

TANKS-STORAGE

Steel; Rubber-Lined Steel; Stainless-Clad Steel; Nickel Clad Steel Stainless Steel; Monel-Clad Steel; Iconell-Clad Steel

INSTALLATION MANHOURS

Capacity Gallons	MANHOURS			
	(1)	(2)	(3)	(4)
20,000	272.0	432.0	360.0	456.0
30,000	320.0	496.0	504.0	520.0
40,000	384.0	576.0	584.0	600.0
50,000	416.0	608.0	616.0	632.0
60,000	448.0	656.0	664.0	680.0
70,000	464.0	720.0	728.0	744.0
80,000	480.0	752.0	760.0	776.0
90,000	504.0	784.0	792.0	808.0
100,000	528.0	800.0	808.0	824.0
200,000	720.0	1064.0	1088.0	1104.0
300,000	816.0	1248.0	1264.0	1280.0
400,000	960.0	1408.0	1416.0	1424.0
500,000	1088.0	1512.0	1516.0	1520.0
600,000	1136.0	1600.0	1600.0	1600.0
700,000	1232.0	1680.0	1680.0	1680.0
800,000	1296.0	1760.0	1760.0	1760.0
900,000	1376.0	1920.0	1920.0	1920.0
1,000,000	1424.0	2080.0	2080.0	2080.0

(1) Steel.

(2) Rubber-lined steel.

(3) Stainless-clad steel; nickel-clad steel.

(4) Stainless steel; monel-clad steel; inconel-clad steel.

Manhours include unloading of shop fabricated sections of sizes within shipping limits, handling, job hauling up to 2000 feet, rigging, picking, setting, job welding, aligning, and checking out of tanks as outlined.

Manhours exclude installation of incoming or outgoing piping and their connections.

For steel cone roof and floating roof tanks increase manhours 20 and 25% respectively.

TANKS-STORAGE

Spheroids; Spheres

INSTALLATION MANHOURS

Capacity Gallons	MANHOURS			
	(1)	(2)	(3)	(4)
10,000	—	608.0	720.0	1088.0
20,000	—	768.0	880.0	1296.0
30,000	—	832.0	1008.0	1456.0
40,000	—	960.0	1120.0	1616.0
50,000	—	1040.0	1216.0	1760.0
60,000	1248.0	1104.0	1280.0	1920.0
70,000	1280.0	1168.0	1338.0	2080.0
80,000	1328.0	1232.0	1624.0	2240.0
90,000	1360.0	1264.0	1456.0	2400.0
100,000	1424.0	1496.0	1520.0	2560.0
200,000	1760.0	1616.0	1648.0	2880.0
300,000	2080.0	—	—	—
400,000	2480.0	—	—	—
500,000	2560.0	—	—	—
600,000	2720.0	—	—	—
700,000	2880.0	—	—	—
800,000	3040.0	—	—	—

(1) Spheroids, steel — 15 psi.

(2) Spheres, steel — 25 psi.

(3) Spheres, steel — 50 psi.

(4) Spheres, steel — 100 psi.

Manhours include unloading of shop fabricated sections of sizes within shipping limits, handling, job hauling up to 2000 feet, rigging, picking, setting, job welding, aligning, and checking out of tanks as outlined.

Manhours exclude installation of incoming or outgoing piping and their connections.

TANKS-STORAGE

Cylindrical

INSTALLATION MANHOURS

Capacity Gallons	Manhours (1)
400	400.0
500	412.0
600	424.0
700	436.0
800	448.0
900	456.0
1,000	464.0
2,000	512.0
3,000	576.0
4,000	608.0
5,000	624.0
6,000	640.0
7,000	648.0
8,000	656.0
9,000	672.0
10,000	688.0
20,000	784.0
30,000	832.0
40,000	896.0
50,000	928.0
60,000	960.0
70,000	976.0
80,000	992.0
90,000	1040.0
100,000	1056.0

(1) Cylindrical, steel — 50-150 psi.

Manhours include unloading of shop fabricated section of sizes within shipping limits, handling, job hauling up to 2000 feet, rigging, picking, setting, job welding, aligning, and checking out of tanks as outlined.

Manhours exclude installation of incoming or outgoing piping and their connections.

For steel cone roof and floating roof tanks increase manhours by 20 and 25% respectively.

TANKS—PROPANE STORAGE

INSTALLATION MANHOURS

Shell Size		Water Cap.	Approx. Weight	Manhours
I.D.	Length	Gallons	Pounds	Each
117"	20'-0"	14,800	33,655	62.4
117"	30'-0"	20,385	46,740	72.0
117"	40'-0"	25,970	59,900	80.4
117"	50'-0"	31,555	73,000	87.6
117"	60'-0"	37,140	86,300	94.8

Propane storage tanks are 250 psi, 1" shell, ½" hemispherical heads, all A-515 pressure vessel quality grade 70 material.

Each tank includes one 18" pad type manway with hinged blind flange, five 3" and four 2"; 300# raised face buttweld flange type nozzles.

Manhours include unloading at job site, job hauling up to 2000 feet, rigging, picking, setting, aligning and checking tanks as outlined.

Manhours exclude installation of incoming or outgoing piping, instrumentation and electrical circuits, and their connections.

COMMINUTORS AND SEWAGE TREATMENT PLANTS—COMMINUTORS

MANHOURS EACH

Drum Diameter Inches	Maximum Capacity Gallons Per Minute	Motor Horsepower	Approximate Weight Pounds	Manhours
5	140	1/3	160	7.4
8	300	1/2	200	9.8
12	850	3/4	325	15.4

SEWAGE TREATMENT PLANTS

MANHOURS EACH

Plant Capacity Gallons Per Day	Tank Size Diameter x Height	Motor Horsepower	Manhours
1,000	5'11' x 7'0''	3/4	32.0
3,000	7'11-1/2'' x 11'4''	1.5	48.0
5,000	9'10'' x 12'0''	2.0	64.0

Manhours include unloading, handling, job hauling up to 2000 feet, rigging, picking, setting, aligning, and checking out of item as listed.

Manhours exclude installation of piping, electrical circuits, and their connections.

Comminutors are motor driven units capable of reducing organic solids in flowing sewage to 1/4-inch size or smaller.

Sewage treatment plants are packaged units designed for purification of liquid sewage by aeration and the reduction of sewage solids by aerobic digestion in the same tank at the same time.

INCINERATORS—LIQUID WASTE

MANHOURS EACH

Item No.	Maximum Waste Flow Pounds Per Hour	Motor Horsepower Required	Required Ground Space Length x Width	Overall Height	Approximate Weight Pounds	Manhours
1	170	5	9'0" x 6'0"	47'0"	7,000	84.0
2	390	10	10'0" x 7'0"	48'0"	9,500	86.0
3	555	15	11'0" x 7'0"	48'0"	13,500	100.0
4	780	20	12'0" x 8'0"	50'0"	17,000	120.0
5	1,000	25	12'0" x 8'0"	51'0"	19,000	128.0
6	1,330	40	13'0" x 9'0"	51'0"	23,500	136.0
7	1,670	50	13'0" x 10'0"	52'0"	29,000	142.0
8	2,650	75	15'0" x 12'0"	53'0"	41,500	160.0
9	1,160	15	9'3" x 5'4"	26'0"	16,000	86.0
10	2,700	25	11'3" x 6'6"	26'0"	23,000	92.0
11	4,620	20 & 7.5	14'2" x 7'7"	26'0"	32,000	95.0

Manhours include unloading, handling, job hauling up to 2000 feet, rigging, picking, setting, aligning, and checkout of items as outlined.

Manhours exclude installation of incoming and outgoing piping, electrical circuits, and their connections.

Items 1 through 8, are factory assembled in three units: the incinerator with its burner and accessories in place, the combustion air blower as a separate assembly, and the stack as a separate assembly.

Items 9, 10, and 11, are factory assembled in two sections: the incinerator with its burner and accessories in place as one assembly and the combustion blower as a separate assembly.

INCINERATORS—SOLID WASTE

MANHOURS EACH

Carburetor Burner BTU Per Hour	Approximate Size of Unit Length x Width x Height	Height of Stack	Weight—Pounds		Manhours
			of Unit	of Stack	
185,000	7'6" x 4'0" x 9'6"	12'0"	12,600	810	72.0
320,000	8'6" x 4'6" x 10'6"	12'0"	18,600	1,035	84.0
570,000	11'6" x 5'0" x 11'6"	12'0"	23,800	1,365	96.0
950,000	13'6" x 6'0" x 12'0"	12'0"	32,200	1,710	108.0
1,500,000	14'6" x 7'0" x 13'6"	12'0"	42,700	2,220	120.0
2,250,000	16'6" x 8'6" x 16'0"	12'0"	63,000	3,060	144.0

Manhours include unloading, handling, job hauling up to 2000 feet, rigging, picking, setting, aligning, and checking out of items as outlined.

Manhours exclude installation of piping, electrical circuits, and their connections.

All items are fully factory assembled, including refractory in two pieces, the incinerator in one piece, and the stack as a separate piece. If additional stack is required this can be added in 4-foot sections.

SOLID WASTE SHREDDERS

Primary, Secondary, and Tertiary Types

MANHOURS EACH   .

Motor Horsepower Range	Approximate Weight Pounds	Capacity Range Tons Per Hour	Manhours
10-25	1,800	1-2	28.0
30-60	6,000	2-5	32.0
60-125	8,000	5-10	32.0
100-200	14,000	10-20	60.0
150-300	16,000	15-30	62.0
200-400	23,000	20-40	128.0
300-600	29,000	30-60	134.0
400-800	48,000	40-80	176.0
800-1,200	105,000	60-80	240.0
1,200-2,000	165,000	80-100	340.0
1,500-3,000	260,000	100-150	480.0

Manhours include unloading, handling, job hauling up to 2000 feet, rigging, picking, setting, aligning, and checking out of shredder of size and capacity as outlined.

Manhours exclude installation of electrical circuits, motors, and their connections.

WASTEWATER TREATMENT PACKAGE SYSTEM

MANHOURS EACH

Capacity Gallons Per Day	Skid Mounted Package Size Length x Width x Height	Total Horsepower	Approximate Weight Pounds	Manhours
7,000	8'0" x 7'0" x 8'8"	2.75	5,700	22.0
15,000	13'0" x 8'0" x 8'8"	2.75	10,000	28.0
25,000	20'0" x 8'0" x 8'8"	3.50	15,000	34.0

Manhours include unloading, handling, job hauling up to 2000 feet, rigging, picking, setting, aligning, and checking out of skid-mounted units as listed.

Manhours exclude installation of incoming, and outgoing piping, electrical circuits, and their connections to and from the skid units.

Items are factory assembled, skid-mounted units consisting of raw water pump, sludge pump, pressure filter feed and backwash pump, pressure filter, surge tank, absorber, absorber aerator, agitators, flocculator clarifier, flash mix tank, belt filter, sludge level sensor, miscellaneous pressure and level instruments, control panel, all tie-in piping, electrical conduit and wiring, and their skid-mounted connections.

AERATORS—MECHANICAL SURFACE TYPE

MANHOURS EACH

Motor Horsepower	Motor RPM	Float Dimensions Diameter x Depth	Approximate Weight Pounds	Manhours
5	1,800	75" x 11"	775	20.0
7.5	1,800	75" x 11"	885	20.0
10	1,200	82" x 13"	930	20.0
15	1,200	82" x 13"	1,350	23.0
20	1,200	82" x 13"	1,450	23.0
25	1,200	94" x 16"	2,020	30.0
30	1,200	94" x 16"	2,220	30.0
40	900	138" x 20"	2,960	40.0
50	900	138" x 20"	3,710	40.0
60	900	138" x 20"	4,060	44.0
75	900	138" x 20"	4,460	44.0
100	900	150" x 20"	6,600	50.0

Manhours include unloading, handling, job hauling up to 2000 feet, rigging, picking, setting, aligning, and checking out of mechanical surface aerators for waste water treatment lagoons or basins.

Manhours exclude installation of piping or electrical circuits and their connections.

ION EXCHANGE DEMINERALIZERS
Two-Bed Type Units

MANHOURS EACH

Tank Size Dia. x Height Inches	Cubic Feet of Cation and Anion	Approximate Unit Weight Pounds	Suggested Flow Rate GPM	Manhours
6 x 66	0.66	645	1.2-1.8	9.6
8 x 66	1.00	717	1.9-3.0	10.0
10 x 66	2.00	870	3.1-4.5	14.4
12 x 72	3.00	1,037	4.6-6.3	17.0
14 x 78	4.00	1,056	6.4-8.5	19.2
16 x 78	5.50	1,414	8.6-12.0	22.0
20 x 78	8.67	1,957	13.0-18.0	28.8
24 x 90	14.50	2,735	19.0-27.0	32.0
30 x 90	22.50	4,107	28.0-40.0	38.4
36 x 90	35.00	5,825	41.0-57.0	42.0
42 x 96	48.00	8,000	58.0-77.0	48.0
48 x 96	60.00	11,000	78.0-100.0	54.0

Manhours include unloading, handling, job hauling up to 2000 feet, rigging, picking, setting, aligning, and checking out of factory prepiped, prewired, fully assembled, skid-mounted, two-bed demineralizer units.

Manhours are for installation of either manual or automatic series units and can be used individually or in series dependent on yield and degree of purity desired.

Manhours exclude installation of piping, electrical circuits, and their connections.

All tanks up to 12 inches in diameter are of P.V.C. construction. All tanks with a diameter of 14 inches or greater are of solid-cast plastisol, lined-steel construction.

ION EXCHANGE DEMINERALIZERS

Mixed-Bed Type Units

MANHOURS EACH

Tank Size Dia. x Height Inches	Cubic Feet of Mixed Resin	Approximate Weight Pounds	Suggested Flow Rate GPM	Manhours
6 x 80	0.50	518	0.7-1.0	14.4
8 x 80	1.00	568	1.1-2.6	15.0
10 x 80	1.67	624	2.7-4.0	21.6
14 x 90	3.50	762	4.1-7.5	25.0
18 x 90	6.00	944	7.6-12.4	33.0
24 x 96	12.00	1,455	12.5-22.0	38.4
30 x 96	19.00	2,117	22.1-34.4	46.0
36 x 102	28.00	2,867	34.5-49.5	50.4
42 x 120	48.00	4,500	49.6-70.0	57.6

Manhours include unloading, handling, job hauling up to 2000 feet, rigging, picking, setting, aligning, and checking out of factory prepiped, prewired, fully assembled, skid-mounted mixed-bed demineralizer units.

Manhours are for installation of either manual or automatic series units.

Manhours exclude installation of piping, electrical circuits and their connections.

All tanks up to 12 inches in diameter are of P.V.C. construction. All tanks with a diameter of 14 inches or greater are of solid-cast plasticol, lined-steel construction.

WATER STILLS

INSTALLATION MANHOURS

Capacity Gallons Per Hour	MANHOURS		
	Steam	Gas	Electrical
1	33.6	33.6	33.6
2	43.2	44.8	49.6
3	49.6	54.4	65.6
4	56.0	62.4	80.0
5	60.8	67.2	92.8
6	64.0	76.8	104.0
7	70.4	81.6	113.6
8	75.2	88.0	126.4
9	78.4	91.2	134.4
10	80.0	96.0	144.0
20	105.6	–	–
30	124.8	–	–

Manhours include unloading, handling, job hauling up to 2000 feet, rigging, picking, setting, aligning, and checking out of items as outlined.

Manhours exclude installation of piping, electrical circuits, and their connections.

Section 2

RELATED EQUIPMENT ITEMS

This section includes manhour tables for fabrication and erection of various items that may be required for or related to a particular piece of equipment.

The following manhour tables are based on averages of many projects installed under varied conditions where strict methods and preplanning were followed and strict reporting of actual time spent was recorded in accordance with the notes as appear on the individual table pages.

The listed manhours include time allowance to complete all necessary labor for the outlined operation.

UNLOADING EQUIPMENT & TANKS
FROM ENCLOSED CARRIER
WITH END OR SIDE OPENING

MANHOURS PER TON

Machinery Classification	MANHOURS		
	Group One	Group Two	Group Three
Lightweight and Bulky			
Up to 1500 pounds	3.50	6.50	4.90
Lightweight and Easily Handled			
Up to 1500 pounds	2.65	3.50	3.35
Heavyweight and Easily Handled			
Up to 10 tons	2.80	—	3.70
Up to 50 tons	2.10	—	3.00
Heavyweight and Bulky			
Up to 10 tons	3.15	—	4.20
Up to 50 tons	2.80	—	3.75

Group one: Using fork truck or other power equipment, drag object from inside to opening and unload to ground or location if adjacent.

Groups two and three: Jacking, bulling, skidding on small rollers, then drag to opening.

Group two: Remove to ground by hand or slide.

Group three: Remove to ground by power equipment.

All groups: Include an allowance for equipment operating crews.

UNLOADING EQUIPMENT & TANKS FROM OPEN CARRIER

MANHOURS PER TON

Machinery Classification	MANHOURS		
	Group One	Group Two	Group Three
Lightweight and Bulky			
Up to 1500 pounds	2.80	3.85	3.15
Lightweight and Easily Handled			
Up to 1500 pounds	2.10	3.15	2.45
Heavy and Easily Handled			
Up to 10 tons	1.75	2.10	1.90
Up to 50 tons	1.05	1.60	1.25
Heavyweight and Bulky			
Up to 10 tons	1.75	2.45	1.95
Up to 50 tons	1.40	2.30	1.75

Group one: Unload to temporary storage adjacent to the carrier.

Group two: Direct to floor location — second or third floor.

Group three: To existing foundation or structural frame work.

All groups: Using forklift truck, derrick, crane or gin pole. Includes allowance for equipment operating crew.

HANDLING AND HAULING
EQUIPMENT AND TANKS
Move Manually With Some or All of the Individual Specified

MANHOURS PER TON

Operation	MACHINERY CLASSIFICATION					
	Lt. Wt. & Bulky	Lt. Wt. & Easily Handled	Heavy Wt. & Easily Handled		Heavy Wt. & Bulky	
	to 1500#	to 1500#	to 10 tons	to 50 tons	to 10 tons	to 50 tons
Jack up & place rollers	0.70	0.56	0.46	0.35	0.53	0.46
Moving on skids or small rollers for 100 feet	1.05	0.98	0.42	0.35	0.60	0.53
Jacking up or down, placing or removing cribbing per ft. of height	0.70	0.56	0.46	0.35	0.53	0.42
Bulling and moving or turning up to 10 feet	1.40	1.23	0.88	0.70	0.95	0.81
Handling cribbing & timber per piece	0.04	0.04	0.04	0.04	0.04	0.04

Manhours are for moving manually with some or all of the individual specified.

HANDLING AND HAULING
EQUIPMENT AND TANKS
Move by Fork Truck, Crane, Hand Truck or Dolly Truck

MANHOURS PER TON

Operation	MACHINERY CLASSIFICATION					
	Lt. Wt. & Bulky	Lt. Wt. & Easily Handled	Heavy Wt. & Easily Handled		Heavy Wt. & Bulky	
	to 1500 #	to 1500#	to 10 tons	to 50 tons	to 10 tons	to 50 tons
Transport for 100 ft. including one lifting operation	0.40	0.35	0.33	0.29	0.38	0.31
Place on base as part of transporting. Does not include line-up	0.35	0.32	0.22	0.18	0.23	0.20
Build up cribbing set object on top prior to lowering or horizontal positioning	0.77	0.63	0.53	0.42	0.60	0.49

Manhours include use of fork lift truck, crane, hand truck and dolly truck.

An allowance for equipment operation crews is included.

ALIGNMENT OF EQUIPMENT

MANHOURS PER WEIGHT UNITS LISTED

Machinery Classification In Pounds	MANHOURS			
	Group One	Group Two	Group Three	Group Four
200 or less	0.60	2.31	3.47	1.40
500	0.74	2.94	4.41	1.40
750	0.81	3.22	4.83	1.40
1000	0.91	3.57	5.36	1.40
1500	1.05	4.20	6.30	1.40
2000	1.23	4.97	6.92	1.40
2500	1.47	5.95	7.65	1.40
3000	1.86	6.30	9.45	1.40
4000	2.63	8.70	10.88	1.40
Per ton above 2	1.40	5.20	7.20	1.40

Group one: Rough alighment if a separate operation. Setting, raising or lowering and removing temporary supporting timbers if used.

Group two: Accurate alignment — pre-assembled at vendors shops, delivered as a single unit. Rough alignment included if combined operation.

Group three: Accurate alignment — disassembled into major sections. Reassembled on frame at location.

Group four: Grouting per square foot.

An allowance for equipment operating crews is included where necessary.

ARC PLATE BUTT WELDING

MANHOURS PER LINEAR FOOT

	PLATE THICKNESS INCHES						
	1/8	3/16	1/4	5/16	3/8	7/16	1/2
Butt Weld							
Flat	—	—	.24	.30	.38	.38	.44
Vertical	—	—	.29	.37	.51	.51	.58
Horizontal	—	—	.35	.42	.64	.64	.71
Overhead	—	—	.37	.45	.77	.77	.84
Flame Cutting	—	.09	.09	.09	.10	.10	.11

	PLATE THICKNESS INCHES						
	9/16	5/8	3/4	7/8	1	1-1/8	1-1/4
Butt Weld							
Flat	—	.58	.70	.77	.83	1.01	1.12
Vertical	—	.84	.88	1.00	1.07	1.31	1.41
Horizontal	—	1.07	1.07	1.12	1.17	1.40	1.49
Overhead	—	1.17	1.17	1.31	1.35	1.76	1.82
Flame Cutting	.11	.12	.13	.16	.17	.18	.20

Manhours include welder and helper time necessary for set-up of machine, procuring welding materials, tackwelding when necessary and welding.

Manhours are based on 100 linear feet or more of welding of the type and size listed. If less than 100 linear feet welding is required, manhours should be increased by at least 25 per cent.

Manhours do not include setting, aligning or positioning of plate or scaffolding. See respective tables for these charges.

ARC PLATE FILLET WELDING

MANHOURS PER LINEAR FOOT

	PLATE THICKNESS INCHES						
	1/8	3/16	1/4	5/16	3/8	7/16	1/2
Fillet Weld							
Flat	.08	.10	.17	.20	.26	.29	.34
Vertical	.10	.14	.20	.27	.34	.36	.42
Horizontal	.11	.15	.18	.32	.39	.40	.46
Overhead	.14	.17	.21	.37	.42	.43	.50
Flame Cutting	—	.09	.09	.09	.10	.10	.11

	PLATE THICKNESS INCHES						
	9/16	5/8	3/4	7/8	1	1-1/8	1-1/4
Fillet Weld							
Flat	.38	.45	.50	.58	.64	—	—
Vertical	.45	.53	.64	.71	.77	—	—
Horizontal	.50	.58	.71	.77	.83	—	—
Overhead	.54	.64	.77	.83	.89	—	—
Flame Cutting	.11	.12	.13	.16	.17	.18	.20

Manhours include welder and helper time necessary for set-up of machine, procuring welding materials, tackwelding when necessary and welding.

Manhours are based on 100 linear feet or more of welding of the type and size listed. If less than 100 linear feet welding is required, manhours should be increased by at least 25 per cent.

Manhours do not include setting, aligning or positioning plate or scaffolding. See respective tables for these charges.

INSULATION OF VESSELS, TANKS AND HEAT EXCHANGERS

MANHOURS PER SQUARE FOOT

Type	Manhours
One layer blocks (wired on)	.048
Additional layer of blocks (wired on)	.032
Sponge felt	.040
Wire mesh	.008
Cement 1/4-in. thick	.024
Cement 1/2-in. thick	.032
Sewed on 8 oz. canvas	.032
Pasted on 8 oz. canvas	.024
Metal lath	.036
Rosin paper	.006
1-in. thick hair felt	.024
1 layer asbestos paper (1/32-in. thick)	.006
Hot pitch or asphalt — one mopping	.007
Two coats of lead and oil paints	.018
Concrete primer and enamel	.034
Cold water paint	.008
Add for weatherproofing with standard covers	.024

Manhours include all operations necessary for the complete installation of the type insulation as outlined above.

Manhours do not include equipment installation or scaffolding. See respective tables for these charges.

MEMBRANE WATERPROOFING VESSELS, TANKS AND HEAT EXCHANGERS

MANHOURS PER SQUARE FOOT

Item	Manhours
One-ply fabric and two moppings	.03
Two-ply fabric and three moppings	.05
Three-ply fabric and four moppings	.06
Four-ply fabric and five moppings	.07
Each additional layer of felt and mopping	.02

Manhours are for waterproofing and dampproofing as itemized and outlined above and include all labor operations necessary for this type of work.

Manhours do not include installation of equipment or scaffolding. See respective tables for these charges.

SUPPORTS FOR FAN AND MOTOR UNITS

MANHOURS PER UNIT

Unit Weight of Fans and Motors	Steel Hangers and Supports in Pounds	MANHOURS		
		Fabricate	Erect	Total
400	150	3.9	3.1	7.0
500	150	3.9	3.1	7.0
600	150	3.9	3.1	7.0
700	200	4.6	3.8	8.4
1300	300	7.7	6.3	14.0
1800	375	9.6	7.9	17.5
2500	375	9.6	7.9	17.5
3900	375	9.6	7.9	17.5
5000	500	11.7	9.0	20.7
6000	500	11.7	9.0	20.7

Manhours include handling, hauling, fabricating and installing hangers and supports for fans and motors outlined above.

Manhours do not include installation of fans or motors.

SUPPORTS FOR
HEATING AND VENTILATING UNITS

MANHOURS PER UNIT

Weight of Unit in Pounds	Steel Hangers and Supports in Pounds	MANHOURS		
		Fabricate	Erect	Total
300	100	3.0	2.6	5.6
450	150	3.9	3.1	7.0
500	200	4.6	3.8	8.4
600	300	7.7	6.3	14.0
900	400	9.6	7.9	17.5
1500	600	13.7	9.1	22.8
1600	650	14.8	9.8	24.6
2500	650	14.8	9.8	24.6
2600	700	15.8	10.6	26.4
3500	700	15.8	10.6	26.4
3700	750	16.9	11.3	28.2
4300	800	18.0	12.0	30.0
4500	800	18.0	12.0	30.0

Manhours include handling, hauling, fabricating and installing supports for heating and ventilating units as outlined above.

Manhours exclude the installation of heating and ventilating units.

SUPPORTS FOR
SELF-CONTAINED AIR-CONDITIONING UNITS

MANHOURS PER UNIT

Refrigeration Tons	Air Conditioning Units in Pounds	Steel Hangers and Supports in Pounds	MANHOURS		
			Fabricate	Erect	Total
6	2000	650	13.5	11.1	24.6
10	3000	700	14.5	11.9	26.4
15	3800	750	15.5	12.7	28.2
20	4000	800	16.5	13.5	30.0
30	4500	900	18.8	15.4	34.2
40	6000	1350	30.6	20.4	51.0
50	8000	1700	38.2	25.4	63.6

Manhours include handling, hauling, fabricating, and installing supports for air conditioni units as outlined above.

Manhours exclude installation of air conditioning units.

SUPPORTS FOR AIR HANDLING UNITS

MANHOURS PER UNIT

Capacity CFM	Steel Hangers and Supports Pounds	MANHOURS		
		Fabrication	Erection	Total
Single Zone Units				
1,000	500	7.9	6.5	14.4
2,500	600	9.9	8.1	18.0
6,000	625	13.2	10.8	24.0
14,000	750	28.8	19.2	48.0
24,000	800	33.8	22.6	56.4
30,000	900	39.6	26.4	66.0
Multi-Zone Units				
4,000	550	8.4	7.3	15.7
6,000	625	13.2	10.8	24.0
10,000	650	15.8	13.0	28.8
15,000	775	31.3	21.4	52.7
22,000	800	33.8	22.6	56.4
30,000	900	39.6	26.4	66.0

Manhours include handling, hauling, fabricating, and installing supports for air handling units as outlined above.

Manhours do not include scaffolding or installation of air handling units. See respective tables for these charges.

DRILLING HOLES IN WELDED ATTACHMENTS

MANHOURS EACH

Thickness of Plate, Angles, etc. Inches	HOLE SIZE IN INCHES		
	3/4 or less	7/8, 1 & 1-1/8	1-1/4, 1-1/2 & 2
1/2 or less	0.2	0.2	0.2
3/4	0.2	0.2	0.3
1	0.2	0.3	0.5
1-1/4	0.3	0.5	0.5
1-1/2	0.5	0.5	0.7
1-3/4	0.5	0.6	0.8
2	0.6	0.8	0.9

Manhours include drilling of hole only.

If holes are to be tapped, increase above manhours 30 per cent.

Drilling of sentinel safety or tell tale holes should be charged at .05 manhours each.

BUTT WELDS-PIPE

MANHOURS EACH

Size Inches	Standard Pipe & OD Sizes 3/8-in. thick	Extra Heavy Pipe & OD Sizes 1/2-in. thick	SCHEDULE NUMBERS								
			20	30	40	60	80	100	120	140	160
1	0.7	0.8			0.7		0.8				1.0
1-1/4	0.8	0.8			0.8		0.8				1.1
1-1/2	0.8	0.9			0.8		0.9				1.3
2	1.0	1.0			1.0		1.0				1.6
2-1/2	1.2	1.3			1.2		1.3				1.8
3	1.3	1.4			1.3		1.4				2.1
3-1/2	1.4	1.6			1.4		1.6				—
4	1.5	1.8			1.5		1.8		2.8		3.0
5	1.7	2.1			1.7		2.1		2.9		3.8
6	2.0	2.5			2.0		2.5		3.8		4.9
8	2.6	3.3	2.6	2.6	2.6	3.0	3.3	4.6	6.0	7.5	8.6
10	3.1	4.0	3.1	3.1	3.1	4.0	5.1	6.8	9.4	11.4	13.1
12	3.6	4.7	3.6	3.6	4.1	5.2	6.6	9.9	12.2	15.3	17.9
14-OD	4.3	5.7	4.3	4.3	5.0	6.8	9.6	13.2	16.2	19.2	22.7
16-OD	5.0	6.6	5.0	5.0	6.6	8.4	12.4	19.5	20.7	25.0	27.7
18-OD	5.9	7.7	5.9	6.8	8.6	11.2	16.4	21.8	25.6	29.9	33.7
20-OD	6.3	8.4	6.3	8.4	9.4	13.8	19.5	26.0	31.9	37.0	40.8
24-OD	6.9	10.1	6.9	—	13.3	20.1	25.2	35.8	43.5	49.3	59.3

Manhours include set-up of welding equipment, welding, grinding where necessary, and stress relieving where necessary.

Stress relieving of welds in carbon steel materials is required by the A.S.A. code for pressure piping, where the wall thickness is 3/4-in. or greater. All the sizes shown below the ruled lines are 3/4-in. or greater.

ERECTION OF BOLT-UPS–PIPE

MANHOURS EACH

Pipe Size Inches	SERVICE PRESSURE RATING					
	150-lb.	300-400-lb.	600-lb.	900-lb.	1500-lb.	2500-lb.
2 or less	0.7	0.8	0.9	1.0	1.2	1.6
2-1/2	0.8	0.9	1.0	1.2	1.5	2.0
3	0.8	0.9	1.0	1.2	1.5	2.0
3-1/2	1.0	1.2	1.3	1.5	1.8	2.4
4	1.2	1.4	1.5	1.7	2.1	2.8
6	1.5	1.7	1.8	2.1	2.6	3.4
8	2.1	2.4	2.6	3.0	3.7	4.9
10	2.7	3.0	3.2	3.7	4.6	6.1
12	3.4	3.8	4.1	4.7	5.8	7.7
14	3.8	4.3	4.6	5.3	6.5	–
16	4.4	4.9	5.2	6.0	7.4	–
18	4.8	5.4	5.8	6.7	8.2	–
20	5.5	6.2	6.6	7.6	9.3	–
24	6.6	7.4	7.9	9.1	11.2	–

Manhours include pick-up of bolts and gaskets at storage and bolting up.

Manhours exclude testing or equipment installation.

MAKE-ON SCREWED FITTINGS–PIPE

MANHOURS EACH

Nominal Size Inches	MANHOURS	
	Plain	Back Welded
1/4	0.1	0.4
3/8	0.1	0.4
1/2	0.1	0.4
3/4	0.1	0.5
1	0.2	0.5
1-1/4	0.2	0.6
1-1/2	0.3	0.7
2	0.3	0.9
2-1/2	0.4	1.0
3	0.4	1.2
3-1/2	0.4	1.4
4	0.5	1.6

Manhours include obtaining fitting from storage and make on at vessel or item required.

Manhours exclude installation of other items or equipment.

HAND EXCAVATION

MANHOURS PER CUBIC YARD

Soil	Excavation	MANHOURS		
		First Lift	Second Lift	Third Lift
Light	general dry	1.07	1.42	1.89
	general wet	1.60	2.13	2.83
	special dry	1.34	1.78	2.37
Medium	general dry	1.60	2.19	2.83
	general wet	2.14	2.85	3.79
	special dry	2.00	2.49	3.31
Hard or heavy	general dry	2.67	3.55	4.72
	general wet	3.21	4.27	5.68
	special dry	2.94	3.91	5.70
Hard pan	general dry	3.74	4.97	6.61
	general wet	4.28	5.69	7.57
	special dry	4.01	5.33	7.09

Manhours include picking and loosening where necessary and placing on bank out of way of excavation, or loading into trucks or wagons for hauling away.

Manhours do not include blasting, hauling or unloading.

Light Soil — Earth which can be shoveled easily and requires no loosening, such as sand.

Medium or Ordinary Soils — Type of earth easily loosened by pick. Preliminary loosening is not required when power excavating equipment such as shovels, dragline scrapers and backhoes are used. This earth is usually classified as ordinary soil and loam.

Heavy or Hard Soil — This type of soil can be loosened by pick but this loosening is sometimes very hard to do. It may be excavated by sturdy power shovels without preliminary loosening. Hard and compacted loam containing gravel, small stones and boulders, stiff clay or compacted gravel are good examples of this type.

Hard Pan or Shale — A soil that has hardened and is very difficult to loosen with picks. Light blasting is often required when excavating with power equipment.

MACHINE EXCAVATION – POWER SHOVEL

MANHOURS PER 100 CUBIC YARDS

Soil	Dipper Size	Manhours
Light	1 cubic yard 3/4 cubic yard 1/2 cubic yard	3.3 4.5 6.0
Medium	1 cubic yard 3/4 cubic yard 1/2 cubic yard	6.0 8.4 11.1
Heavy	1 cubic yard 3/4 cubic yard 1/2 cubic yard	8.1 11.1 14.7
Hard pan	1 cubic yard 3/4 cubic yard 1/2 cubic yard	10.2 13.8 18.3
Rock	1 cubic yard 3/4 cubic yard 1/2 cubic yard	10.2 13.8 18.3

Manhours include operations of swamping and excavating and dumping on sidelines or into trucks.

If excavations are to be greater in depth than 6 feet, the estimator should consider additional methods, planning and equipment required.

Manhours do not include hauling or blasting.

Light Soil – Earth which can be shoveled easily and requires no loosening, such as sand.

Medium or Ordinary Soils – Type of earth easily loosened by pick. Preliminary loosening is not required when power excavating equipment such as shovels, dragline scrapers and backhoes are used. This earth is usually classified as ordinary soil and loam.

Heavy or Hard Soil – This type of soil can be loosened by pick but this loosening is sometimes very hard to do. It may be excavated by sturdy power shovels without preliminary loosening. Hard and compacted loan containing gravel, small stones and boulders, stiff clay or compacted gravel are good examples of this type.

Hard Pan or Shale – A soil that has hardened and is very difficult to loosen with picks. Light blasting is often required when excavating with power equipment.

Rock – Requires blasting before removal and transporting. (May be divided into different grades such as hard, soft, or medium.)

MACHINE & HAND BACKFILL

Average for Sand or Loam, Ordinary Soil, Heavy Soil, and Clay

MANHOURS PER UNITS LISTED

Item	Unit	Manhours
Hand place	cu. yd.	0.55
Bulldoze loose material	100 cu. yds.	3.32
Clamshell:		
1 cu. yd. bucket	100 cu. yds.	3.20
3/4 cu. yd. bucket	100 cu. yds.	4.00
1/2 cu. yd. bucket	100 cu. yds.	5.50
Hand Spread:		
Stone or gravel	cu. yd.	0.40
Sand	cu. yd.	0.35
Cinder	cu. yd.	0.40
Tamp by hand	cu. yd.	0.60
Pneumatic Tamping	cu. yd.	0.25

Hand place manhours include hand shoveling of loose earth within throwing distance of stock piles. This unit does not include compaction.

Bulldoze loose material manhours include the moving of pre-stockpiled loose earth over an area.

Clamshell manhour units include the placement of materials from reachable stockpiles.

Stone, sand and cinder manhour units include the hand shovel placing of these materials from strategically located stockpiles.

Tamp by hand and pneumatic tamping manhours include the compacting of pre-spread materials in 6-in. layers.

Above manhours do not include trucking or fine grading.

DISPOSAL OF EXCAVATED MATERIALS

MANHOURS PER HUNDRED (100) CUBIC YARDS

Truck Capacity and Length of Haul	Manhours								
	Average Speed 10 mph			Average Speed 15 mph			Average Speed 20 mph		
	Truck Driver	Laborer	Total	Truck Driver	Laborer	Total	Truck Driver	Laborer	Total
3 Cu Yd Truck:									
1 Mile Haul	15.0	2.8	17.8	11.6	2.8	14.4	10.5	2.8	13.3
2 Mile Haul	21.8	2.8	24.6	16.2	2.8	19.0	14.0	2.8	16.8
3 Mile Haul	28.2	3.0	31.2	20.6	3.0	23.6	17.3	3.0	20.3
4 Mile Haul	36.0	3.0	39.0	26.8	3.0	29.8	21.0	3.0	24.0
5 Mile Haul	41.7	2.5	44.2	31.00	2.5	33.5	25.5	2.5	28.0
4 Cu Yd Truck:									
1 Mile Haul	11.3	2.1	13.4	8.8	2.0	10.8	7.9	2.1	9.0
2 Mile Haul	16.2	2.1	18.3	12.0	2.0	14.0	10.4	2.1	12.5
3 Mile Haul	21.6	2.0	23.6	15.8	2.3	18.1	13.2	2.2	15.4
4 Mile Haul	26.4	2.0	28.4	18.7	2.3	21.0	15.6	2.2	17.8
5 Mile Haul	31.3	1.3	32.6	22.2	1.6	23.8	18.5	1.5	20.0
5 Cu Yd Truck:									
1 Mile Haul	9.0	1.7	10.7	7.0	1.7	8.7	6.3	1.6	7.9
2 Mile Haul	13.0	1.7	14.7	9.7	1.7	11.4	8.3	1.7	10.0
3 Mile Haul	17.1	1.8	18.9	12.3	1.8	14.1	10.4	1.7	12.1
4 Mile Haul	21.0	2.0	23.0	15.0	2.0	17.0	12.4	1.7	14.1
5 Mile Haul	25.0	1.7	26.7	17.9	1.7	19.6	14.8	1.6	16.4
8 Cu Yd Truck:									
1 Mile Haul	5.6	1.0	6.6	4.8	1.0	5.8	4.0	1.0	5.0
2 Mile Haul	8.2	1.0	9.2	6.0	1.0	7.0	5.2	1.0	6.2
3 Mile Haul	10.5	1.1	11.6	7.8	1.1	8.9	6.5	1.0	7.5
4 Mile Haul	13.2	1.1	14.3	9.2	1.1	10.3	7.6	1.0	8.6
5 Mile Haul	15.6	1.3	16.9	10.9	1.3	12.2	9.0	1.1	10.1

Manhours include round trip for truck driver, spotting at both ends, unloading and labor for minor repairs.

Manhours do not include labor for excavation or loading of trucks. See respective tables for these charges.

WOOD FORMS FOR EQUIPMENT FOUNDATIONS
SIMPLE LAYOUT

MANHOURS PER SQUARE FOOT

Item	Manhours
Square Pads 6-in. to 18-in. High Ground Floor	
Build in place	0.18
Strip and clean	0.04
Total	0.22
Square Pads 6-in. to 18-in High Elev. Floors	
Build in place	0.20
Strip and clean	0.05
Total	0.25
Square Pads to 4-ft. High Ground Floor	
Fabricate and erect	0.22
Strip and Clean	0.06
Total	0.28
Square Pads to 4-ft. High Elev. Floors	
Fabricate and erect	0.24
Strip and clean	0.07
Total	0.31

Manhours are based on the fabrication and installation of 2-in. materials for formwork to 18-in. high, and plywood sheathing for forms to 4-ft. high. All properly braced and anchored in place.

A simple layout is that of a small square pad poured either integral with floor or over pre-set dowels left purposely in pre-poured floor for this reason.

Manhours do not include the placement or setting of anchor bolts or miscellaneous embedded steel items. See respective tables for these charges.

WOOD FORMS FOR EQUIPMENT FOUNDATIONS COMPLEX LAYOUT

Bulky, Offset, Skewed, and Angled

MANHOURS PER SQUARE FOOT

Item	Manhours
Average All Heights and Sizes	
Fabricate and erect	0.30
Strip and clean	0.17
Total	0.47
Tank Cradle Forms	
Build in place	0.19
Strip and clean	0.04
Total	0.23

Complex foundation manhours are average for all sizes and shapes and are based on the use of 1- and 2-in. planking, plywood sheathing and minor sheet metal cuts and bends.

A complex layout is that of a large and bulky foundation with many offsets, skews and angles, such as a foundation for a turbo-generator, etc.

Manhours do not include the placement or setting of anchor bolts or miscellaneous embedded steel items. See respective tables for these charges.

REINFORCING RODS AND MESH

MANHOURS PER UNITS LISTED

Item	MANHOURS	
	Per Ton	Per CWT
Unload, sort and pile rods	1.75	0.0875
Fabricate — Cut and Bend		
1/2-in. round and larger	6.00	0.3000
3/8-in. round and smaller	11.48	0.5740
Place Loose Without Tieing		
3/4-in. round and larger	7.25	0.3625
5/8-in. round and smaller	8.78	0.4390
Place and Tie Rods		
Walls, columns, etc.	16.50	0.8250
Floors	22.10	1.1050
Average All Operations — All Sizes		
Without tieing	18.51	0.9250
With tieing	29.79	1.4900

Item	Manhours per 100 sq. ft.	
Welded Wire Mesh Cut and place	.80	—

Manhours include all necessary handling, hauling, fabricating and installation as required for the above described items.

ANCHOR BOLTS–HOOK TYPE

0′ 8″ – 4′ 0″

INSTALLATION MANHOURS EACH

Size Inches	Manhours Each for Over-all Length							
	0' 8"	1' 0"	1' 6"	2' 0"	2' 6"	3' 0"	3' 6"	4' 0"
1/4	.15	.15	.20	.20	–	–	–	–
3/8	.15	.15	.20	.22	–	–	–	–
1/2	.15	.15	.25	.28	–	–	–	–
5/8	.15	.20	.25	.28	.30	.33	–	–
3/4	.18	.20	.28	.30	.35	.38	–	–
7/8	–	–	.40	.43	.45	.48	.50	.53
1	–	–	.40	.45	.48	.50	.53	.58
1-1/4	–	–	.48	.50	.50	.53	.55	.65
1-1/2	–	–	.50	.55	.55	.58	.60	.70
1-3/4	–	–	.55	.58	.60	.65	.68	.73
2	–	–	–	.65	.68	.70	.75	.78
2-1/4	–	–	–	.70	.73	.75	.78	.80
2-1/2	–	–	–	.75	.78	.78	.80	.85

Manhours are based on over-all length of anchor bolt from end to end including hook. They include installation of template and bolt, or bolt and sleeve, as the case may be.

All bolts 7/8-in. and larger are assumed to be sleeved and those smaller than 7/8-in. round are assumed to be without sleeves.

Manhours exclude fabrication of bolts.

ANCHOR BOLTS–HOOK TYPE

4′ 6″ – 8′ 0″

INSTALLATION MANHOURS EACH

Size Inches	Manhours Each for Over-all Length							
	4' 6"	5' 0"	5' 6"	6' 0"	6' 6"	7' 0"	7' 6"	8' 0"
7/8	.90	.98	1.10	1.15	1.23	1.30	1.30	1.40
1	.93	1.00	1.15	1.25	1.28	1,38	1.35	1.50
1-1/4	.95	1.10	1.25	1.28	1.30	1.40	1.45	1.54
1-1/2	.98	1.15	1.28	1.33	1.38	1.43	1.48	1.58
1-3/4	1.00	1.25	1.33	1.40	1.43	1.50	1.55	1.60
2	1.10	1.28	1.40	1.45	1.50	1.58	1.60	1.65
2-1/4	1.15	1.33	1.45	1.48	1.55	1.60	1.63	1.68
2-1/2	1.25	1.40	1.48	1.53	1.60	1.63	1.68	1.70

Manhours are based on over-all length of anchor bolt from end to end including hook. They include installation of bolt and sleeve.

Manhours exclude fabrication of bolt.

INSTALLATION OF STRAIGHT TYPE ANCHOR BOLTS

MANHOURS EACH

Size	Manhours Each for Overall Length					
	0' 8"	1' 0"	1' 6"	2' 0"	2' 6"	3' 0"
1/4"	.10	.12	.15	.18	.20	.25
3/8"	.10	.12	.18	.20	.22	.28
1/2"	.10	.15	.18	.20	.25	.30
5/8"	.15	.15	.20	.22	.28	.33
3/4"	.15	.18	.20	.25	.30	.38
1"	.15	.18	.23	.25	.33	.40

LOOPS AND SCREW ANCHORS

MANHOURS EACH

Item	Manhours
Coil Loops—½"x4" through 1½"x12"	0.15
Screw anchors—½" through 1½"	0.12

Manhours are based on the installation of template and bolt for the size and length as outlined above and are average for all heights.

If a mixed crew of various crafts is used in the setting of above bolts, consideration should be given this when arriving at a composite rate for the conversion of manhours to labor dollars.

Manhours do not include engineering time spent in the aligning or checking of bolts. This is usually a part of field overhead and should be considered as such.

For sizes not listed, take the next highest listing.

MISCELLANEOUS FASTENERS

HOURS REQUIRED EACH OR PER OPERATION

Bolt Size Inches	Lead Expansion Anchors	Toggle Bolts	Wooden and Lag Screws	Mach. Screws In Steel Drill & Tap	Mach. Bolt in Steel Av. 3/8" Thick
1/8"	–	.11	–	.32	.24
3/16"	.14	.12	–	.39	.27
1/4"	.15	.14	–	.42	.30
5/16"	.18	–	–	.48	.34
3/8"	.25	.18	–	.58	.38
7/16"	.28	–	–	.65	.44
1/2"	.28	–	–	–	–
5/8"	.38	–	–	–	–
#10 x 1"	–	–	.03	–	–
#12 x 1-1/4"	–	–	.03	–	–
1/4" x 1-1/2"	–	–	.05	–	–
3/8" x 2"	–	–	.08	–	–
1/2" x 2-1/2"	–	–	.11	–	–

Manhours include checking out of job storage, handling, hauling, fabricating hole with power tool when required, and erection of anchor bolt or screw.

Manhours are average for heights to 25 feet.

Manhours exclude straps or installation of other supports. See respective tables for these time requirements.

HANGERS AND FASTENERS

HOURS REQUIRED PER HUNDRED

Size Inches	One Hole Strap Type Height to				Split Pipe Rings & Sockets Height to				Pipe Riser Clamps Height to			
	10'	15'	20'	25'	10'	15'	20'	25'	10'	15'	20'	25'
3/8	1.37	1.40	1.44	1.47	–	–	–	–	–	–	–	–
1/2	1.37	1.40	1.44	1.47	–	–	–	–	–	–	–	–
3/4	1.37	1.40	1.44	1.47	–	–	–	–	–	–	–	–
1	1.86	1.90	1.95	2.00	–	–	–	–	–	–	–	–
1-1/4	2.74	2.80	2.88	2.94	17.46	18.00	18.55	18.90	48.01	49.50	51.00	51.93
1-1/2	2.74	2.80	2.88	2.94	26.19	27.00	27.80	28.35	53.84	55.50	57.15	58.28
2	4.12	4.20	4.33	4.41	33.46	34.50	35.55	36.20	58.20	60.00	61.80	63.00
2-1/2	4.12	4.20	4.33	4.41	40.74	42.00	43.25	44.10	62.57	64.50	66.95	67.72
3	6.86	7.00	7.21	7.35	48.00	49.50	51.00	51.98	63.39	70.50	72.60	74.00
3-1/2	8.43	8.60	8.86	9.03	62.56	64.50	68.45	67.70	77.12	79.50	81.92	83.47
4	9.80	10.00	10.30	10.50	69.85	72.00	74.15	75.60	87.30	90.00	92.70	94.50

Size Inches	Item Description	Height to			
		10'	15'	20'	25'
–	Beam Clamps	26.30	27.00	27.80	28.35
1/4	Rod-Size Expansion Anchors	29.30	30.00	30.90	31.50
3/8	Rod-Size Expansion Anchors	47.50	48.00	49.45	50.40
1/2	Rod-Size Expansion Anchors	56.50	57.00	58.70	59.85
–	Concrete Inserts 3/8" or 1/2" Nuts	36.75	37.50	38.65	39.40
–	Ceiling Flanges and Sockets	29.30	30.00	30.90	31.50

Manhours include checking out of job storage, handling, hauling, and installing items as outlined.

Manhours exclude installation of electrical devies. See respective tables for these time frames.

ANGLES, SLEEVES, INSERTS, SLOTS & UNISTRUT

MANHOURS PER UNITS LISTED

Item	Unit	Manhours
Pipe Sleeves Through 12-in. Dia.		
Substructure	each	.620
Super Structure	each	.730
Pipe Sleeves Through 36-in. Dia.		
Substructure	each	.780
Super Structure	each	.900
Inserts		
Substructure	each	.121
Superstructure	each	.152
Anchor Slots & Unistrut		
Substructure	lin. ft.	.011
Superstructure	lin. ft.	.018
Embedded Angle		
Substructure	pound	.021
Superstructure	pound	.033

Manhours are average for the above described items and included handling, hauling and installing.

Manhours do not include fabrication.

CONCRETE FOR EQUIPMENT FOUNDATIONS

MANHOURS PER CUBIC YARD

Item	Manhours
Square Pads	
Crane & bucket	1.88
Crane, bucket & buggies	2.50
Offset, Skewed and Angles	
Crane & bucket	3.20
Crane, bucket & buggies	4.25

Manhours are for the placement and vibration of concrete for the above items.

Square pad manhours are based on pouring of square pads to 4 feet high either integral with floor or over pre-set dowels on pre-poured floor.

Offset, skewed or angled manhours are based on that of pouring a large and bulky foundation with offsets or angles or both.

Manhours do not include forming, placing of reinforcing steel, or concrete finishing. See respective tables for these charges.

CONCRETE FINISH

MANHOURS PER SQUARE FOOT

Surface Finish	Manhours
Carborundum rub	.045
Remove fins or ties — point & patch	.030
Machine trowel & hand burnish	.015
Hand steel trowel	.030
Wood float	.001
Broom	.003
Screeding off	.006
Cure & protect	.002
Grout 1-in. thick	.500

Manhours are for the above types of finish and include all necessary operations as may be required.

Manhours do not include the placement of concrete or concrete items. See respective tables for these charges.

CONCRETE TOPPING FINISH

MANHOURS PER SQUARE FOOT

Item	Manhours
Integral Topping	
1/2-in. by hand	.029
1/2-in. by machine	.018
1-in. by hand	.068
1-in. by machine	.022
Separate Topping	
1/2-in. by hand	.036
1/2-in. by machine	.024
1-in. by hand	.072
1-in. by machine	.028
1-1/2 - in. by hand	.080

Manhours are for topping finish and include mixing and stand-by time where required.

Manhours do not include pouring of foundations. See respective table for this charge.

Section 3

TECHNICAL INFORMATION

This manual is solely intended for the estimation of labor and not for the design of systems or items. Therefore, this section has been held to a minimum and includes only information that will benefit the estimator in the preparation of an estimate.

This section contains a table showing the conversion of minutes to decimal hours, a manhour table for the installation of patent scaffolding, and several weight tables showing the weights of various steel and metal sheets.

ERECT AND DISMANTLE
PATENT SCAFFOLDING

MANHOURS PER SECTION

Length	One or Two Sections High			More than Two Sections High		
	Erect	Dismantle	Total	Erect	Dismantle	Total
One to two sections long	1.4	1.0	2.4	1.7	1.2	2.9
Three to five sections long	0.9	0.6	1.5	1.0	0.7	1.7
Six or more sections long	0.7	0.4	1.1	0.9	0.5	1.4

Patent tubular steel scaffolding consisting of sections 7-ft. long x 5-ft. wide x 5-ft. high with 2-in. planking top.

Manhours include handling and hauling scaffolding and materials from and to storage, erection, leveling, securing and dismantling of scaffolding and scaffolding materials.

MINUTES TO DECIMAL HOURS
CONVERSION TABLE

Minutes	Hours	Minutes	Hours
1	.017	31	.517
2	.034	32	.534
3	.050	33	.550
4	.067	34	.567
5	.084	35	.584
6	.100	36	.600
7	.117	37	.617
8	.135	38	.634
9	.150	39	.650
10	.167	40	.667
11	.184	41	.684
12	.200	42	.700
13	.217	43	.717
14	.232	44	.734
15	.250	45	.750
16	.267	46	.767
17	.284	47	.784
18	.300	48	.800
19	.317	49	.817
20	.334	50	.834
21	.350	51	.850
22	.368	52	.867
23	.384	53	.884
24	.400	54	.900
25	.417	55	.917
26	.434	56	.934
27	.450	57	.950
28	.467	58	.967
29	.484	59	.984
30	.500	60	1.000

WEIGHT TABLE—GALVANIZED STEEL SHEET

Gauge	Pounds Per Square Foot
10	5.781
11	5.156
12	4.531
14	3.281
16	2.656
18	2.156
20	1.656
22	1.406
24	1.156
26	.906
27	.844
28	.781
30	.656

WEIGHT TABLE—STAINLESS STEEL SHEET

Thickness In Inches	U.S. Std. Gauge	Pounds Per Square Foot
.1406	10	5.906
.125	11	5.25
.1093	12	4.593
.0937	13	3.937
.0781	14	3.281
.0625	16	2.625
.050	18	2.10
.040	—	1.648
.0375	20	1.575
.035	—	1.442
.0312	22	1.3125
.025	24	1.05
.020	—	.824
.0187	26	.7875
.016	—	.659

WEIGHT TABLE—ALUMINUM SHEET & PLATE

B & S Gauge Number	Thickness (In Inches)	Approximate Weight Per Square Foot	
		Sheet	Plate
—	* .190	2.68	—
—	.188	2.65	—
—	* .160	2.26	—
—	.156	2.20	—
—	.125	1.76	—
10	.102	1.44	—
—	* .100	1.41	—
11	.091	1.28	—
—	* .090	1.27	—
12	.081	1.14	—
—	* .080	1.13	—
13	.072	1.04	—
—	* .071	1.00	—
14	.064	.903	—
—	* .063	.889	—
16	.051	.716	—
—	* .050	.706	—
18	.040	.568	—
20	.032	.450	—
22	.025	.357	—
24	.020	.283	—
26	.016	.225	—
28	.012	.178	—
30	.010	.141	—
32	.008	.113	—
34	.006	.085	—
—	2.000	—	28.2
—	1.750	—	24.7
—	1.500	—	21.2
—	1.250	—	17.6
—	1.000	—	14.1
—	.875	—	12.3
—	.750	—	10.6
—	.625	—	8.8
—	.500	—	7.1
—	.375	—	5.28
—	.313	—	4.40
—	.250	—	3.52

* American Standard Preferred Thickness.

WEIGHT TABLE—BRASS SHEET

Thickness in Inches	B & S Gauge	Pounds Per Square Foot	Thickness in Inches	B & S Gauge	Pounds Per Square Foot
1.000	—	44.06	.0571	15	2.516
.875	—	38.56	.0508	16	2.238
.750	—	33.05	.0453	17	1.996
.625	—	27.54	.0403	18	1.776
.500	—	22.03	.0359	19	1.582
.4600	4/0	20.27	.0320	20	1.410
.4096	3/0	18.05	.0285	21	1.256
.375	—	16.52	.0253	22	1.115
.3648	2/0	16.07	.0226	23	.9958
.3249	1/0	14.32	.0201	24	.8857
.3125	—	13.77	.0179	25	.7887
.2893	1	12.75	.0159	26	.7006
.2576	2	11.35	.0142	27	.6257
.250	—	11.02	.0126	28	.5552
.2294	3	10.11	.0113	29	.4979
.2043	4	9.002	.0100	30	.4406
.1875	—	8.262	.0089	31	.3922
.1819	5	8.015	.0080	32	.3528
.1620	6	7.138	.0071	33	.3129
.1443	7	6.358	.0063	34	.2776
.1285	8	5.662	.0056	35	.2468
.125	—	5.508	.0050	36	.2203
.1144	9	5.041	.0045	37	.1983
.1019	10	4.490	.0040	38	.1763
.0907	11	3.997	.0035	39	.1542
.0808	12	3.560	.0031	40	.1366
.0720	13	3.173	.0028	41	.1234
.0641	14	2.825	.0025	42	.1101

WEIGHT TABLE—COPPER SHEET

Thickness In Inches	Nearest B & S Gauge	Pounds Per Square Foot	Thickness In Inches	Nearest B & S Gauge	Pounds Per Square Foot
.3451	2/0	16	.0755	13	3.5
.3235	1/0	15	.0647	14	3
.3019	1	14	.0593	15	2.75
.2804	1	13	.0539	16	2.5
.2588	2	12	.0485	16	2.25
.2372	3	11	.0431	17	2
.2157	4	10	.0377	19	1.75
.2049	4	9.5	.0323	20	1.5
.1941	4	9	.0270	21	1.25
.1833	5	8.5	.0243	22	1.13
.1725	5	8	.0216	23	1
.1617	6	7.5	.0189	25	0.87
.1510	7	7	.0162	26	0.75
.1402	7	6.5	.0135	27	0.63
.1294	8	6	.0108	29	0.5
.1186	9	5.5	.0081	32	0.37
.1078	10	5	.0054	35	0.25
.0970	10	4.5	.0027	41	0.13
.0863	11	4			

JOB ESTIMATING FORM

SHEET NO ___ OF ___

ESTIMATE NO.

COMPANY				COMPOSITE CREW RATE		SHEET NO
PROJECT		LOCATION				
DESCRIPTION OF WORK			ESTIMATOR	CHECKED BY	DATE IN	DATE DUE

No.	Description	Unit	Quantity	Weight		Unit Man-Hours	Total Man-Hours	Unit Labor Cost	Unit Material Cost	Total Cost		
				Unit	Total					Labor	Material	Total

Gulf Publishing Company, Houston   Form 310

This Job Estimating Form is ideal for use when working with the Estimating Man-Hour Manuals.

Printed and bound by CPI Group (UK) Ltd, Croydon, CR0 4YY

08/05/2025

01864834-0001